KU-165-284

NELSON BLACKIE

MATHEMATICS IN ACTION

Mathematics in Action Group

Cambridge House Boys' Grammar School
CAMBRIDGE AVENUE
BALLYMENA

1

Thomas Nelson and Sons Ltd
Nelson House Mayfield Road
Walton-on-Thames Surrey
KT12 5PL UK

51 York Place
Edinburgh
EH1 3JD UK

Nelson Blackie
Westercleddens Road
Bishopbriggs
Glasgow
G64 2NZ UK

Thomas Nelson (Hong Kong) Ltd
Toppan Building 10/F
22A Westlands Road
Quarry Bay Hong Kong

Thomas Nelson Australia
102 Dodds Street
South Melbourne
Victoria 3205 Australia

Nelson Canada
1120 Birchmount Road
Scarborough Ontario
M1K 5G4 Canada

© Mathematics in Action Group 1986

First published by Blackie and Son Ltd 1986
ISBN 0-216-91907-x

This edition published by Thomas Nelson and Sons Ltd 1992

ISBN 0-17-431410-8
NPN 9 8 7 6 5 4 3 2

All rights reserved. No paragraph of this publication may
be reproduced, copied or transmitted save with written
permission or in accordance with the provisions of the
Copyright, Design and Patents Act 1988, or under the
terms of any licence permitting limited copying issued
by the Copyright Licensing Agency, 90 Tottenham Court
Road, London, W1P 9HE.

Any person who does any unauthorised act in relation to
this publication may be liable to criminal prosecution and
civil claims for damages.

Printed in Great Britain.

Robin D. Howat, Auchenharvie Academy, Stevenston, Ayrshire
Edward C. K. Mullan, Eastwood High School, Glasgow
Ken Nisbet, Madras College, St Andrews, Fife
Doug Brown, St Anne's High School, Heaton Chapel, Stockport, Cheshire

with

**W. Brodie,   D. Donald,   E. K. Henderson,   J. L. Hodge,   J. Hunter,   R. McKendrick,
H. C. Murdoch,   A. G. Robertson,   J. A. Walker,   P. Whyte,   H. S. Wylie**

MATHEMATICS IN ACTION GROUP

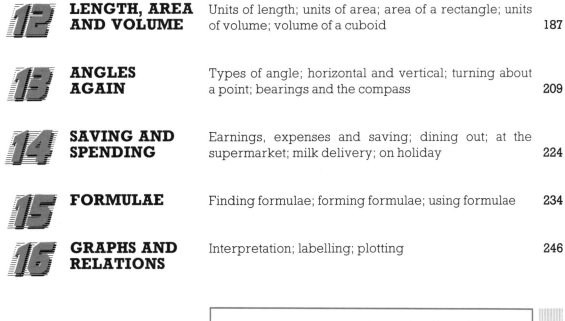
INVESTIGATIONS will always be in a right-handed box

BRAINSTORMERS will always be in a left-handed box

Mathematics is in action all around you. You need mathematics in your daily life. **Mathematics in Action** has been written to help you to understand and to use mathematics sensibly and well—to save you time and effort. Some parts of mathematics are needed in other subjects and some parts will be studied in greater detail later on.

Mathematics in Action follows the latest thinking in *what* mathematics should be studied and how hard, or easy, it should be. So you are taken forward, stage by stage, as far as you can go.

Exercises for practice, Puzzles and Games for fun, Brainstormers to make you think, Investigations to explore, Practical Activities, even Check-ups (to see how you're doing)—all are here.

Enjoy maths with **Mathematics in Action!** Let's hope that you will find a lot that is worthwhile, interesting, and above all, useful.

MiAG April 1986

INSTRUCTIONS AND ORDERS

All these leaflets give you *instructions*.

What do the *titles* tell you to do?

Imagine having to follow all the instructions before you could find out what they were for.

A **flowchart** is a list of instructions that have to be carried out in a certain order. It always has a title that tells you what the instructions are for.

Picture story

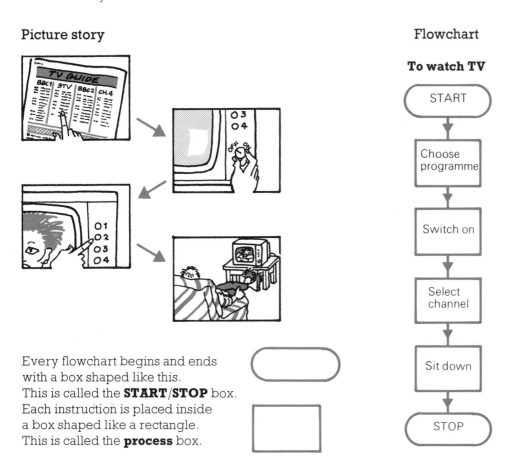

Every flowchart begins and ends with a box shaped like this. This is called the **START/STOP** box. Each instruction is placed inside a box shaped like a rectangle. This is called the **process** box.

Flowchart

To watch TV

START

Choose programme

Switch on

Select channel

Sit down

STOP

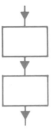

Boxes are joined by **flowlines**.
You have to follow the direction of the arrows.

 You may find it easier to draw flowcharts on squared paper than on plain paper.

=== *Exercise 1A* ===

1 How to play a tape in a cassette recorder. Draw a flowchart with these four instructions. Remember the START and STOP boxes, and the title.

(i) Choose a tape.

(ii) Place in recorder.

(iii) Push 'Play' button.

(iv) Listen.

2 The 'Fireworks Code'. Draw a flowchart using these instructions. Remember to give it a title.

(i) Choose a firework.

(ii) Close the box.

(iii) Place on firm ground.

(iv) Light the blue paper.

(v) Stand well back.

(vi) Enjoy it safely.

3 Getting ready for school. Draw a flowchart for this picture story.

 Get up
 Wash
 Dress
 Eat breakfast
 Go to school

The order in which you carry out instructions is important.

Why is this person puzzled?

Exercise 1B

1 Sort these instructions into the proper order, then put them in a flowchart. Remember to give it a title.

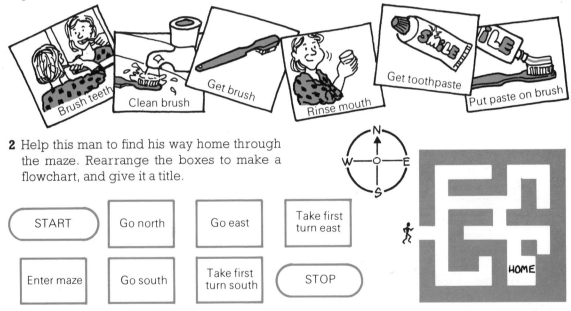

2 Help this man to find his way home through the maze. Rearrange the boxes to make a flowchart, and give it a title.

START   Go north   Go east   Take first turn east

Enter maze   Go south   Take first turn south   STOP

3 I started with the number 5 and ended with the number 3.
I used a flowchart made up of the following boxes:

STOP   ÷ 3   × 2   − 1   START

Draw the flowchart I used.

4 Draw a flowchart to help Tom to get to the third floor.

===== *Exercise 1C* =====

1 Recipes in older cookery books give oven temperatures in degrees Fahrenheit. Modern cookers use temperatures in degrees Celsius. Here is a flowchart to convert °F to °C.

Fahrenheit to Celsius

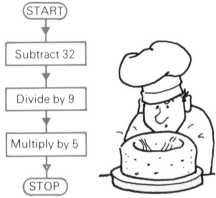

START

↓

Subtract 32

↓

Divide by 9

↓

Multiply by 5

↓

STOP

Use the flowchart to change these temperatures to °C.
a 32°F   **b** 77°F   **c** 140°F   **d** 212°F.

2 a Make a flowchart to change Celsius to Fahrenheit.   *HINT*   Reverse everything.
  b How can you check that your flowchart works?
  c Now check, using 100°C.

3 Here is a flow chart for estimating the depth of a well in metres.

How deep is the well?

To estimate depth in metres

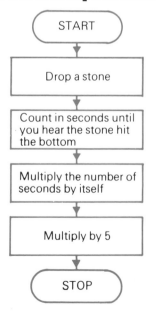

START

↓

Drop a stone

↓

Count in seconds until you hear the stone hit the bottom

↓

Multiply the number of seconds by itself

↓

Multiply by 5

↓

STOP

How deep will these three wells be if you hear the stone hit the bottom after:
a 3 seconds   **b** 5 seconds   **c** 8 seconds?

THE DECISION BOX AND BRANCHES

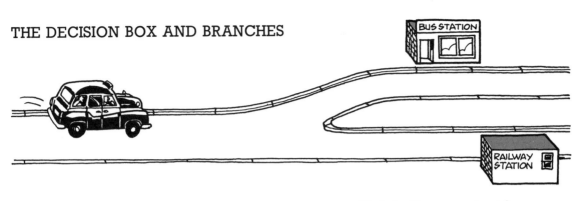

The passenger in the taxi has a choice: he can take either the bus or the train. Look at the taxi-driver's problem in the form of a flowchart.

The diamond-shaped **decision** box has two exits, labelled 'Yes' and 'No'.

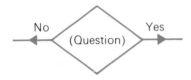

In this case, after the decision is made, the problem **branches**.

To take the correct road

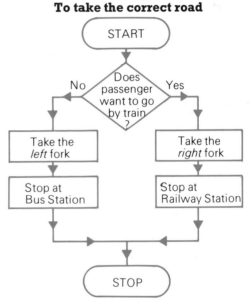

═══════════════ *Exercise 2A* ═══════════════

1 You cannot make up your mind whether to watch a western on Channel 1 or a cartoon on Channel 2. So you toss a coin. Heads—Channel 1, Tails—Channel 2.
Arrange the boxes to make a flowchart for this.

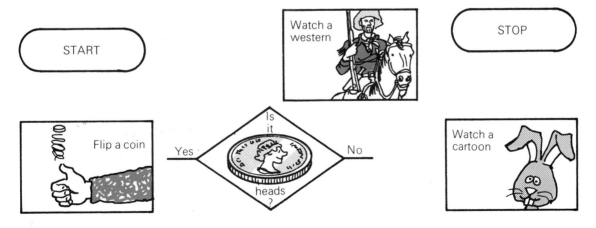

2 Jim is delivering a parcel to a block of flats. If the flat is on the first or second floor he will use the stairs. If it is above the second floor he will use the lift.
Arrange the boxes in the form of a flowchart.

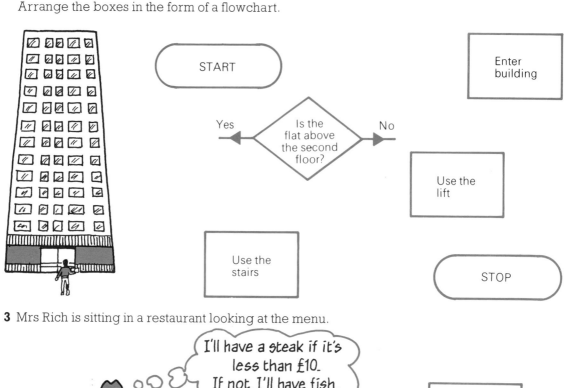

3 Mrs Rich is sitting in a restaurant looking at the menu.

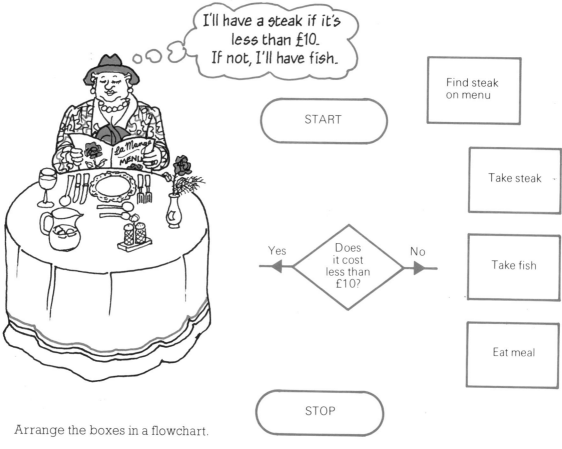

Arrange the boxes in a flowchart.

4 Ann has saved some of her pocket money to go to the cinema. She is 14 years old and so is allowed to watch only 'PG' films. If the film starts after 7 pm she must be accompanied by an adult.

Look at the flowchart. What is the missing question in the third decision box?

Copy and complete the flowchart.

Use it to find out whether Ann can go to the cinema to see:

a a PG film, at 4.30 pm, on her own

b a PG film, at 7.30 pm, on her own

c a PG film, at 8 pm, with an adult

d a film which is not rated PG, at 5 pm, on her own.

Going to the cinema

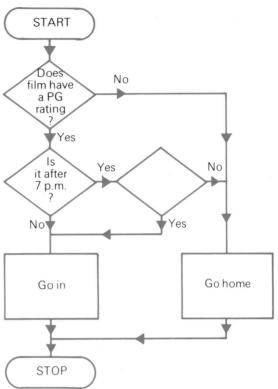

═══════════════ *Exercise 2B* ═══════════════

1 Sunny Jim, staying on a camping site, decides to shop at the local village the next day. He thinks: 'If it's raining I'll take the car'. 'If I'm walking, will I have time to take the longer scenic route?'

Arrange his thoughts in a flowchart using these boxes.

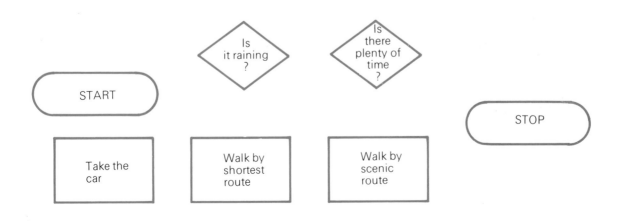

FLOWCHARTS

2 The underground railway in Glasgow has an inner circle and an outer circle. There are 15 stations.

A passenger is waiting at station 1.

If she wants to go to station 2, 3, 4, 5, 6 or 7 she uses the inner circle.

For station 10, 11, 12, 13, 14 or 15 she uses the outer circle.

For station 8 or 9 she takes the first train to arrive.

Mary, Anne and Linda are waiting at station 1.

Mary is going to station 14, Anne to station 4 and Linda to station 8.

Arrange these boxes in a flowchart. Check that it works for all three girls.

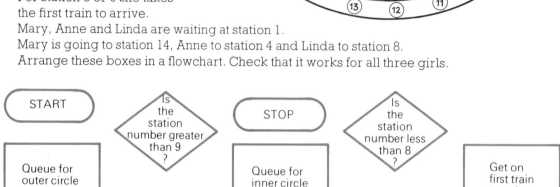

3 Some pupils are helping to plan a weekend trip to the school cottage.

Tom: I'm sure no more than 12 will want to go. A minibus will be big enough.

Ann: What if more than 12 *do* want to go?

Tom: If there are 24, or less, two minibuses will do.

Jill: What if the whole class of 30 decide to go?

Tom: Then we'll need a coach.

Use these boxes to draw a flowchart which describes their plans.

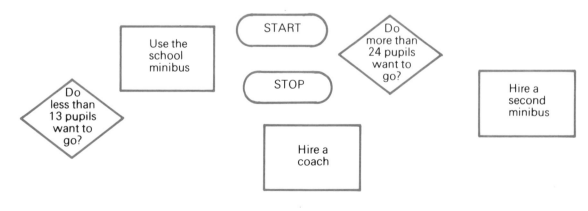

THE DECISION BOX AND LOOPS

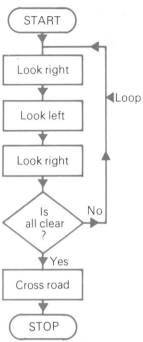

Do you remember the 'Green Cross Code', or 'How to cross a road safely'?

The road is very busy, so you would have to keep looking until it was clear.

Again there is a **decision** box in the flowchart, but this time the problem does not branch.

If 'all is not clear' you have to repeat the process of checking the road both ways.

The problem **loops** back to the beginning until 'all *is* clear'.

=== *Exercise 3A* ===

1 It's Peter's turn to play darts. Use the boxes to draw his flowchart. (Don't forget to choose a title.)

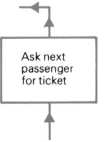

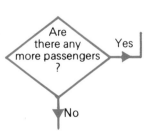

2 The ticket inspector calls 'Have your tickets ready, please!'
Arrange the boxes to make the **All tickets please** flowchart.

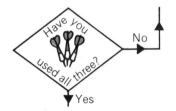

3 $5 \times 3 = ?$

This is an 'empty' flowchart.
Here are the boxes for it.

Five threes

START

STOP

Have you been here 5 times ? — No

Yes

Begin with a total of zero

Add 3 to total

Make a note of your total

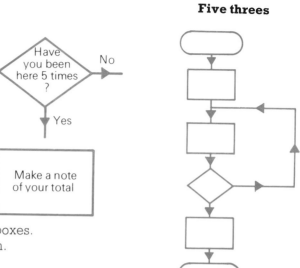

Copy the flowchart and then fill in the boxes.
Use your flowchart to do the calculation.

4 $3 \times 5 = ?$

Draw a flowchart for this calculation.

─────────────────────── *Exercise 3B* ───────────────────────

1 Copy this empty flowchart, then fill in the boxes to show Iain how to win a prize.

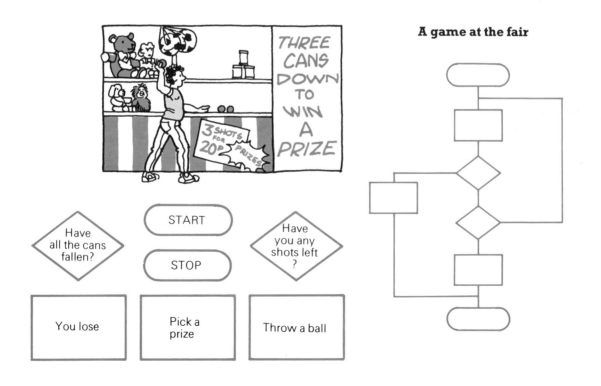

THREE CANS DOWN TO WIN A PRIZE

3 SHOTS FOR 20P PRIZES

A game at the fair

Have all the cans fallen?

START

STOP

Have you any shots left ?

You lose

Pick a prize

Throw a ball

2 $15 \div 3 = ?$ **Fifteen divided by three**

Copy this empty flowchart, and use the boxes to fill it in.

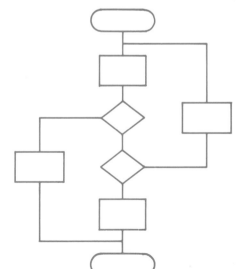

| Begin with 15 |

| Subtract 3 |

| Is the result zero ? |

| Keep a note of how often you pass here |

| Answer is the number of times you went round the loop |

(START)

(STOP)

3 Why does a similar flowchart not work for $3 \div 15 = ?$

4 Roll the dice

Here are the scoring rules of the game. Always throw two dice.

a If you throw 2—disaster! Wipe out! This and all previous scores reduced to zero. Turn finished.

b If you throw 7—add to previous total. Turn finished.

c If you throw any other score, add it on to previous total and have another throw.

Copy the empty flowchart, and use the boxes to fill it in to describe the rules of the game.

a

b

c   Add 5 to score so far

| Throw dice | | Total score zero |

| Add score to total | | Does throw total 2? |

(START)

(STOP)

| Add score to total | | Does throw total 7? |

THE READ/WRITE BOX

To calculate an average

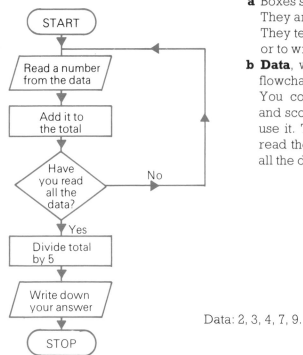

This flowchart has two new features:

a Boxes shaped like this.
They are called **read**/**write** boxes.
They tell you to read information
or to write down the answer.

b Data, which is listed at the bottom of the
flowchart.
You copy the data into your notebook,
and score out each number in turn as you
use it. The loop takes you back round to
read the next number in the data list until
all the data has been used.

Data: 2, 3, 4, 7, 9.

═══════════════════ *Exercise 4A* ═══════════════════

1 a Follow the flowchart above, and so calculate the average of the five numbers listed in the
data.
b Try it again using data: 4, 7, 10, 12, 17.

2 A teacher reads out the register for the pupils' attendance.
Copy the empty flowchart and use the boxes below to
fill it in. Remember to give it a title.

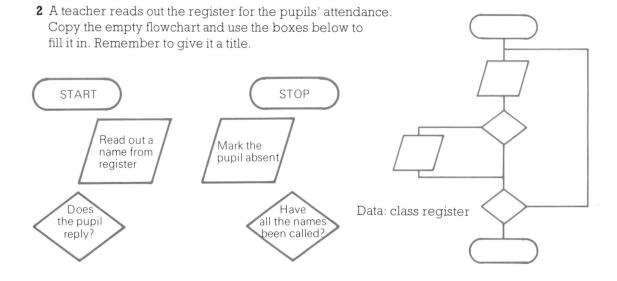

Data: class register

3 To decode a message you need a **KEY**. Here is the key for this code.

KEY

SPACE	A	B	C	D	E	F	G	H	I	O	U	P
0	1	2	3	4	5	6	7	8	9	10	11	12

Follow the flowchart to decode this message:

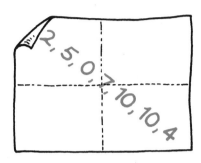

Data: 2, 5, 0, 7, 10, 10, 4

To decode a message

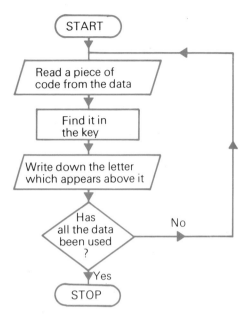

4 Here is a flowchart whose title has been deliberately missed out.

   a Use the flowchart to process the data.

   b Examine your answers and suggest a title for the flow chart.

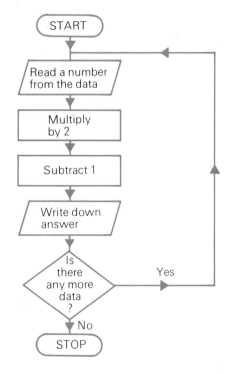

Data: 1, 2, 3, 4, 5, 6, 7, 8, 9, 10.

5 Using the same data as question **4**, make one change in the flowchart so that it produces even numbers.

FLOWCHARTS

1

Use the flowchart and data to work out how much the girl spends at the sweet shop.

2 A number trick

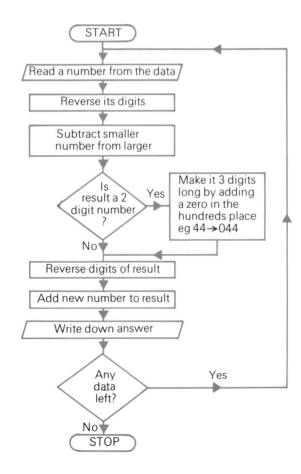

To work out a bill

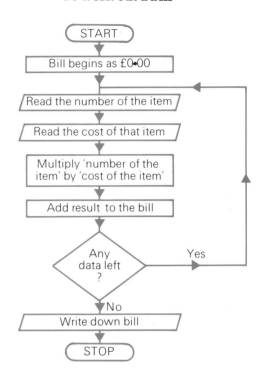

Data: 2, 10p, 3, 2p, 5, 1p.

a Process the data.
b What have you found?
c Try it again with your own data, consisting of 3-digit numbers in which the first digits are different from the last ones.

Data: 432, 257, 372, 892, 754.

Here is a flowchart without a title.

1 Process the data using the flowchart.

2 Describe the purpose of the flowchart.

3 Suggest a title.

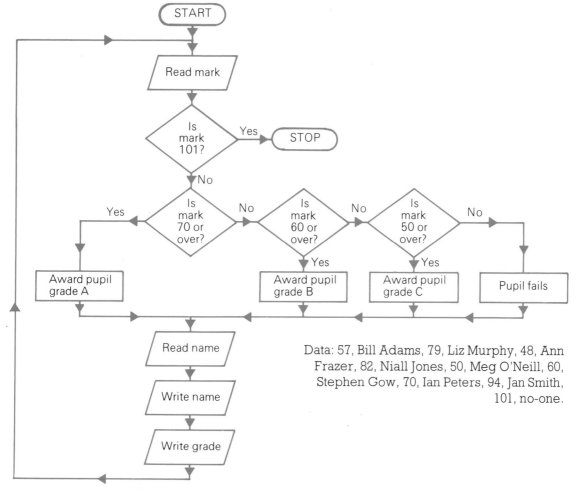

Data: 57, Bill Adams, 79, Liz Murphy, 48, Ann Frazer, 82, Niall Jones, 50, Meg O'Neill, 60, Stephen Gow, 70, Ian Peters, 94, Jan Smith, 101, no-one.

4 Which piece of data isn't used?

5 Suggest why a mark of 101 is used to get out of the loop.

6 It has been decided that a mark in the range 40–49 will be given a grade D, and 30–39 a grade E.
Redraw the flowchart to include these options.

CHECK-UP ON **FLOWCHARTS**

1A Copy this flowchart, with all the dotted lines and boxes, into your notebook.

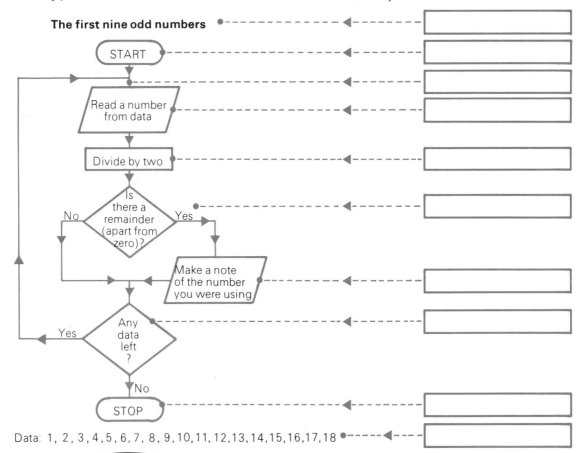

2A a Fill in the boxes in your diagram, using these labels: Title, START box, read box, write box, decision box, flowline, process box, data list, decision box answer, STOP box.

b Draw the decision box which starts a **branch** in the flowchart.

c Draw the decision box which controls a **loop** in the flowchart.

d Process the data using the flowchart.

3B, C A friend who is visiting you has never used a telephone. You have to explain how to do this.

Draw a flowchart which shows the steps in making a telephone call.

0 1 2 3 4 5 10 67 348 2121 1000 001

AT THE FAIRGROUND

The boys and girls from Brightwell Youth Club have arrived in Sunsea for a day's outing to the fairground.

First, some of them tried to win a prize at the 'Roll-a-ball' stall.

=== *Exercise 1A   Roll-a-ball* ===

1 To win a prize a player must score
15 or more with two shots.
What did each boy or girl score?
Who won a prize?

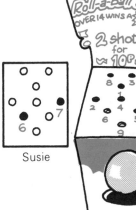

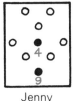 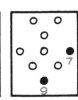 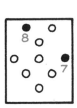

Jenny   Peter   James   Irene   Sita   Susie

2 a At another Roll-a-ball stall three shots cost 15p.
To win a prize a player must score 20 with three shots.
What did each of these score? Did anyone win a prize?

Emma   Ian   Salim   Neil   Kate   Geoff

b Here are the scores of six members of the club.

Fiona 7, 8	Maureen 5, 9	Alec 5, 7
Jim 9, 7	David 4, 9	Elizabeth 3, 8

How much does each one need to score with their third shot to win a prize.

c Tom, Tracy and Colin each have one more ball to roll after scoring as follows:

Tom   3, 5   Tracy   8, 9   Colin   7, 6

Can they all win a prize? Give a reason for your answer.

3 Make a list of all the ways in which you
could score exactly 22 with three shots.

22 POINTS WITH 3 SHOTS WINS THE JACKPOT!

=============== Exercise 1B   Throw-a-ring ===============

1 Lisa, John and Kevin took turns
at throwing 4 rings.

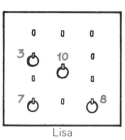

Lisa   John   Kevin

Find their scores.
Does it matter in which order you add up the numbers each time?

2

	Points
Alan	2, 7, 8, 11
Ben	3, 9, 10, 11
Charles	5, 10, 0, 8
David	6, 7, 8, 7
Ewan	8, 8, 9, 7
Fred	4, 11, 11, 4

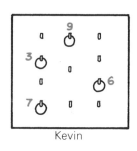

30 OR MORE WINS A PRIZE!

Who won a prize?
What do you think happened to Charles' third shot?

3 A score of exactly 25 wins a special jackpot prize.
Here are some scores after three throws.
Alison 6, 6, 6   Barbara 4, 8, 10   Carol 9, 7, 5
Donna 4, 4, 8   Elaine   11, 6, 3   Fiona 6, 0, 8
How many points would each girl need to score with her fourth throw to win the jackpot?

4 Here are some more scores.

Susie   7, 4, 9   Tom   6, 0, 10   Irene 11, 5, 11

They each have one more ring to throw.
a Can they all still win the jackpot?
b Can they all still win an ordinary prize?
Explain your answers clearly.

5 What is the highest possible score with four rings?

6 List all the ways you can score a total of 7.
All the rings must stay on the board!

When it started to rain, some members went to the games room to play darts.

Exercise 1C   Darts

1 Caroline, Jo and Sue decided to throw nine darts each. The one with the highest score would be the winner.

Caroline	Jo	Sue
5	17	16
12	3	0
18	14	17
20	20	5
1	19	9
3	7	54
24	15	6
19	20	11
25	13	13

What was the winning total? Who won?
How did Sue manage to score 54 with one of her darts?

2 Jason and Kevin each start with 301 points.
They subtract each score for three darts from their points. The first to reach zero wins.

Here are their scores: Copy and complete their calculations:

Jason	Kevin
70	68
51	94
66	45
51	49
23	40
38	5

Jason	Kevin
301	301
-70	-68
231	233
-51	-94
180	

Who won the game? By how many points?

3 Jenny and Lisa also start with 301 points each.
After they have thrown five darts each, their scores are:
Jenny 42, 63, 81, 17, 73
Lisa 72, 58, 65, 54, 28
How many points do they each still need to win the game?

4 Peter and Geoff start with 501 points each.
After six throws, Peter has scored 62, 55, 77, 49, 8 and 36,
and Geoff has scored 71, 39, 101, 18, 42 and 56.
How many more points does each require to win the game?

ON TOUR

Exercise 2A

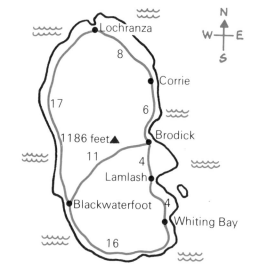

This map shows the Isle of
Arran in the Firth of Clyde.
The numbers give the
distances in miles between
the villages.
A party of schoolchildren arrive
by boat at Brodick pier for a
camping and hiking holiday on the island.

1 The children decide to camp for the
night at Blackwaterfoot.
  a How many routes are marked from Brodick
  to Blackwaterfoot?
  b List the villages on each route and the total
  distance involved.
  c The shortest route might not be the easiest
  for them. Why not?

2 The children spend their second night at Lochranza. Next day they will hike to Whiting Bay.
How far is the journey:
  a down the west coast;
  b down the east coast?
How many miles would be saved
by taking the shorter route?

3 The trip finished with a bus tour right round the island. What distance was this?

4 This mileage chart shows the distances between five towns. For example, by reading down and across, you can see that the distance from Northport to Eastport is 51 miles. Copy the map, and mark in the distances between the towns.

Capital city				
28	North-port			
38	66	South-port		
29	51	42	East-port	
31	35	43	60	West-port

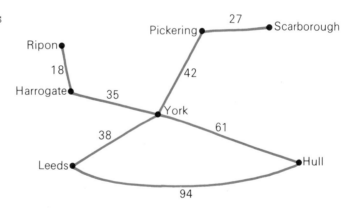

 Make up a mileage chart like the one shown above for the distances between the villages on Arran. Take the shortest distance in each case.

Exercise 2B

Distances are shown in kilometres on this road map.

1 The Wilson family set off by car from Leeds to Scarborough, travelling via York.

  a How many kilometres will their journey be?

  b After 70 kilometres they stop for a picnic. How far do they still have to go?

Ripon● 18 Harrogate● 35 ●York 42 Pickering● 27 ●Scarborough 61 ●Hull 38 Leeds● 94

2 A bus travelling from Ripon to York breaks down after 29 kilometres.

  a How far has the bus still to go?

  b The bus driver telephones a garage in the nearest town. Which town?

3 Shahid plans to travel from Leeds to Hull. He is not sure which route to take. He can choose a direct route or one which takes him through York.
Which route is shorter, and by how much?

4 a What is the shortest distance by road from Hull to Ripon?

  b A car travelling from Hull to Ripon on this route has a puncture after 58 kilometres. How far has it still to go?

5 Jean Stevens lives in Hull and Anna Peters lives in Scarborough. They both support York City football team. On Saturday they are both travelling to York to see a match. Who has further to travel? How much further?

6 Mr Williams is travelling from Reading to Swindon. Between junctions 11 and 12 on the M4 motorway he hears on his car radio that an overturned lorry is blocking the motorway between junctions 12 and 13. He must leave the motorway.

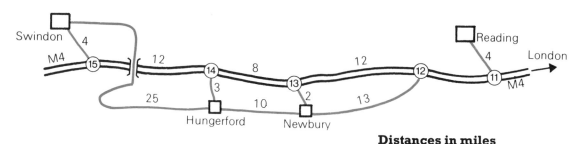

Distances in miles

  a What then is the shortest route to Swindon?
  b How far is it from junction 12 to Swindon by this route?
  c Why might he decide to take a longer route?

AT THE SCHOOL SHOP

=========================== *Exercise 3A* ===========================

1 Which is the cheapest item on sale at the shop? Which is the dearest one?

2 If you had 50p to spend at the school shop, what would you buy? How much change should you be given?

3 How much would it cost to buy one of each item in the picture? What change would you be given from a £1 coin?

4 Look at this table. How much did each pupil spend at the shop?

	Allan	Mark	Isobel	Omar	Anita	Lorna	Cathy	David
Crisps			1		1	2		
Chocbar		1		1			1	
Chewy				1			1	1
Apple			1			1	1	
Fruito	1							1

What change would each receive from 50p?

5 If you spent 30p, which two items did you buy?

6 Tom has £1.25 to spend, but this has to last him the whole week. He buys crisps and an apple on Monday, an apple and a Chocbar on Tuesday, Fruito and an apple on Wednesday, and a packet of Chewy on Thursday.
Has he enough money left to buy anything on Friday?
If so, what can he buy? How much change would he have left?

============================== *Exercise 3B* ==============================

1 Mr Sharp runs the school shop. At the beginning of the week he buys from the 'Fair Deal' Cash and Carry:

> 10 boxes of crisps (50 packets per box)
> 280 Chocbars, 300 packets of Chewy,
> 12 dozen apples, 425 tubs of Fruito.

The sales during that week were:

	Monday	Tuesday	Wednesday	Thursday	Friday
Crisps	84	73	62	93	113
Chocbar	57	42	38	69	57
Chewy	63	54	42	56	71
Apple	36	28	18	24	13
Fruito	26	43	34	128	54

a Calculate how many of each item are left at the end of the week.

b On which day were most items sold? How many?

c On which day were fewest items sold? How many?

2 If Mr Sharp had bought the goods from the 'Cut Price' Cash and Carry every item, including each packet of crisps and each apple, would have cost him 1p less. How much would he have saved altogether?

3 The crisps Mr Sharp bought were:
Plain—3 boxes, Cheese and Onion—3 boxes,
Chicken—2 boxes, Salt and Vinegar—2 boxes.
At the end of the week he was left with the following number of packets:
Plain—18, Cheese and Onion—27, Chicken—17, Salt and Vinegar—13.
How many of each flavour had been sold?

=============================== *Exercise 3C* ===============================

1 Read Mr Sharp's shop rules. Why do you think he made them?

2 Without breaking the shop rules, show how you would pay for:

a 2 Chocbars	**b** 1 apple	**c** 5 packets of crisps
d 5 Fruitos	**e** 3 Chocbars	**f** 1 Chocbar and 1 apple
g 2 apples	**h** half a dozen apples	**i** 10 packets of Chewy

3 If you keep to the rules there is one item you cannot buy on its own. Which one? What is the smallest number of this item you would have to buy?

4 How much would the following pay?
Fiona—2 apples and a packet of crisps
Iqbal—4 packets of Chewy
Nina—2 packets of crisps and a Chocbar
Rory—A Chocbar and a packet of crisps.
Who would not be served because the shop rules would be broken?

MAGIC SQUARES

==== *Exercise 4* ====

1 a Add up the numbers in each row of the magic square above.
  b Add up the numbers in each column.
  c Add up the numbers in each diagonal.
    What do you notice about all of these?

2 Copy and complete these boxes to make them magic squares.

2		
	5	
4		8

8	1	6
	5	

		4
	5	
6		2

2		
	7	
6	1	

3 Make up a '3 by 3' magic square of your own.
Use each number from 1 to 9 once only, and put 5 in the middle.
Make sure that the numbers in each row, column and diagonal add up to 15.

4 Try to make up a '3 by 3' magic square that does not have 5 in the middle square. Write a sentence about the result.

5 Copy and complete the following to make magic squares.

28	21	
	25	
	29	

2	14		
	3		6
	5		4
8	12	1	

21			17	9
2	19	6		
8	25			16
	1	18	10	
20	7	24	11	3

6 Copy this on to squared paper. Use your calculator to complete it as a magic square.

1	99	3			5	94	8	92	10
90				86	85	17	83	19	11
80	79	23	77		26	74	28	22	
31	69	68	34	66	65	37	33	62	
60		58		45	46	44	53	49	51
	52		47	55	56		48		41
61	32		64	36		67	63	39	70
21	29	73	27	75		24	78	72	30
20	82	18	84	15	16	87	13	89	
91	9	93	4	6	95	7	98		100

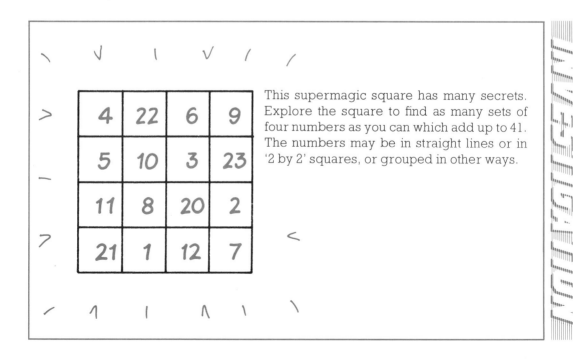

This supermagic square has many secrets. Explore the square to find as many sets of four numbers as you can which add up to 41. The numbers may be in straight lines or in '2 by 2' squares, or grouped in other ways.

4	22	6	9
5	10	3	23
11	8	20	2
21	1	12	7

PUZZLES

Cross-numbers
Copy this puzzle on to squared paper and complete it.

Clues Across
1 A century plus 15.
3 Days in a week plus days in January.
5 Emergency phone number minus 283.
7 17 + 17 + 17 + 17.
8 Sum of first 6 odd numbers.
9 Number of dots on a dice.
10 Number of weeks in 3 years.
12 Days in a leap year minus days in a week.
14 Sides of a triangle plus sides of an octagon.
15 Degrees in a right angle.
16 Battle of Hastings year.
18 All fingers and toes.

Clues Down
2 A book's pages are numbered 7 to 63. How many pages?
3 817 − 449.
4 Total pence in one of each kind of British coin.
6 1000 + 900 + 60 + 6.
7 Think of a number. Add 1000. Take away 379. Take away number first thought of.
8 3 dozen plus 3.
9 Add four 7s.
10 3 half centuries.
11 Letters in the alphabet minus the vowels.
12 Seconds in $6\frac{1}{2}$ minutes.
13 Baker's dozen plus 7.
14 A gross plus a score.
17 Change from a £1 when 37p is spent.

Star gazers
Find the numbers hiding behind the stars. Then make up a similar question for your neighbour.

a 65	**b** 5*	**c** *3	**d** 75	**e** 5*	**f** 23*
+ **	+ *3	+ 7*	− **	− 28	− *23
* 04	* 28	* 00	27	*8	8

27

SEQUENCES OF NUMBERS

A list of numbers such as 2, 4, 6, 8, . . . is called a *sequence of numbers*. For many sequences there is a rule which helps you to find the next number in the sequence.
A possible rule for the sequence 2, 4, 6, 8, . . . is *Add 2.*

========================= *Exercise 5A* =========================

1 Think of the house numbers in the picture above: 2, 4, 6, . . .
   What will the numbers of the houses across the road be?
   Write down a rule for finding the next number in this sequence.

2 Write down a rule for each of the following sequences.
   Use your rule to write down the next number in each sequence.
   a 2, 5, 8, 11, . . .   **b** 1, 6, 11, 16 . . .   **c** 10, 9, 8, 7, . . .   **d** 20, 40, 60, . . .
   e 100, 95, 90, . . .   **f** 101, 102, 103, . . .   **g** 7, 16, 25 . . .   **h** 46, 39, 32, . . .

3 List the number of dots in each pattern.
   Then write down the next two numbers in the sequence.

   a   **b**

   c   **d**

4 What is the missing number in each of these sequences?
   Describe the rule you have used for each one.
   a 15, 20, __, 30   **b** 12, __, 8, 6   **c** 3, 6, __, 12
   d 60, 48, 36, __   **e** 1, 10, 19, __, 37   **f** __, 111, 222, 333

5 Copy this pattern of numbers, which is
   called Pascal's triangle.
   Make the triangle larger by writing
   down the next three rows. Describe the
   rule that you have used.

   ```
           1
          1  1
         1  2  1
        1  3  3  1
       1  4  6  4  1
   ```

6 Make up some sequences of your own, and write down the first four numbers in each of them.
   Ask your neighbour to find the rule you used for each one.

1 Write down two more numbers in each of these sequences:

a 0, 2, 4, 6, ... **b** 5, 10, 15, ... **c** 1, 11, 21, ...

d 1, 2, 4, 7, 11, ... **e** 1, 3, 7, 13, ... **f** 7, 11, 13, 17, 19, ...

2 What rule can you find for each of these sequences?

a 9, 27, 81, 243 **b** 1, 2, 4, 8, 16 **c** 1000, 100, 10, 1

d 1, 5, 25, 125 **e** 64, 16, 4, 1 **f** 108, 36, 12, 4

3 A number is missing in each sequence. Can you work out the one missing?

a 13, 17, 21, __, 29 **b** 300, 200, 100, __ **c** 77, 88, 99, __, 121

d 11, 111, __, 11111 **e** 1, 2, 3, 1, 2, 4, 1, 2, __ **f** 1, 2, 4, 7, 11, 16, __, 29

g 99, 87, 75, 63, __ **h** 55, 47, 40, 34, 29, __ **i** 1, 3, 6, 10, __, 21

4 Find four different rules for sequences beginning 1, 2, ...
Write down the next three numbers in each sequence.

5 Look at the way in which numbers are put in the circles below:
Copy and complete each diagram.

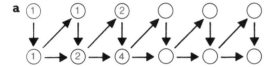 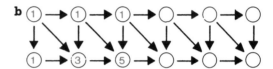

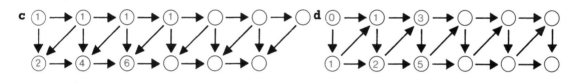

6 Write down *all* the numbers in the circles in question **5d**, in order of size, beginning 0, 1, 1, 2, 3, 5, ...
Try to find a simple rule which will give you a number in this sequence.
This is called a Fibonacci sequence. Leonardo Fibonacci discovered it in the thirteenth century. He was the despair of his teachers, but grew up to use mathematics to help people.

OUT AND ABOUT WITH NUMBERS

Exercise 6A

1 The first hole at the Royal Troon golf course is 343 yards long. Tom Watson drove 285 yards from the tee straight up the fairway. How far was his ball then from the hole?

2 A computer salesman kept a note of the amount of petrol he bought in six months.

Month	January	February	March	April	May	June
Litres of petrol	141	138	157	149	156	143

How much petrol did he buy altogether?

3 In a block of flats a milkman delivered 4, 6, 3, 4, 5, 9, 2 and 6 bottles of milk. How many bottles did he take from his van?

4

The number of miles travelled by a car by the beginning and end of one year are shown on the windscreens. How many miles did the car travel during the year?

5 A lorry driver travels 119 miles from Aberdeen to Edinburgh and then 373 miles to London. How long was his whole journey?

6 The measurements of this L-shaped metal plate are in centimetres. Calculate the total distance round its edges.

7 Two friends, Joe and Tom, were looking for 50 sponsors altogether for a school mini-marathon. Joe found 17 and Tom found 19. How many more sponsors did they need?

8 The attendances at three football matches were 25 067, 32 741 and 19 558. What was the total attendance? How many short of 100 000 is this?

9 The Greatbrit Building Society received £1 205 497 from savers, and lent £958 076 to home owners. How much money did the society keep in reserve?

10 a How many days are there in the first six months of this year?

b How many days are there in the second six months? Check your answer by calculating it in a different way.

Exercise 6B

---

1 A building firm employed 20 bricklayers, 5 joiners, 4 plumbers, 4 electricians, 6 painters and 2 foremen during May. In June 7 bricklayers were paid off, but 2 more joiners and a security guard were taken on. How many people were employed in June?

2 A new garage opened on the first of August. That day there were three deliveries of petrol. This table shows the amounts in litres.

	2 Star	3 Star	4 Star
1st delivery	4250	3570	6450
2nd delivery	3250	2850	9750
3rd delivery	1975	2460	8870

During the month the garage sold 5876 litres of 2 Star, 4077 litres of 3 Star and 18 908 litres of 4 Star petrol.

How many litres of petrol were left in the tanks at the end of August?

3 The lengths on this metal plate are measured in millimetres.

Are you told enough to calculate the lengths marked A and B?

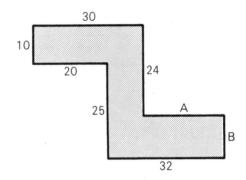

4 The P.E. department at Eglinton Academy agreed to organise an Easter skiing trip to Switzerland. The cost of the trip would be £147, and this had to be paid by March.

a By March, Bob had only made two payments of £25 each. How much had he still to pay?
b Jill paid £15 in September, £32 in October, £23 in November, £8 in December, £14 in January and £37 in February. How much must she pay in March?

5 Find a tenth number for the sequence 1, 4, 16, 64, ... Calculate the sum of the first ten numbers in the sequence.

6 Six athletes took part in the final of the 'Mr Superfit' contest. All six contestants had to do four types of exercise, each lasting for one minute. The athlete with the highest overall total score would be the winner.

	Squat thrusts	Pressups	Bar jumps	Situps
Bobby Biceps	56	63	72	84
Hammy String	61	54	70	85
Nobby Knees	59	58	76	77
Philip Phlab	48	36	83	63
Harry Hulk	57	68	73	79
Peter Pectoral	52	56	75	83

Find their total scores and the order in which they finished.

MAKING SURE OF ADDITION AND SUBTRACTION

NUMBERS IN ACTION—1

Mr Speedie urgently needed answers to the calculations below, but his calculator had **broken down**!

So he set all the calculations down neatly on paper, and worked out the answers himself.

He managed to do Exercise 7A correctly in 8 minutes. He had only one mistake in Exercise 7B, which took him 10 minutes.

See if you can beat him.

Exercise 7A

1	13 +65	**2**	46 +22	**3**	33 +55	**4**	25 +54	**5**	61 +37	**6**	57 +32
7	152 +345	**8**	507 +282	**9**	194 +704	**10**	64 +26	**11**	18 +15	**12**	39 +43
13	36 +76	**14**	58 +85	**15**	137 +257	**16**	256 +348	**17**	592 +239	**18**	456 +654
19	58 −24	**20**	36 −21	**21**	67 −35	**22**	74 −63	**23**	98 −54	**24**	175 −65
25	765 −352	**26**	897 −564	**27**	75 −29	**28**	76 −28	**29**	42 −15	**30**	51 −36
31	92 −29	**32**	51 −24	**33**	33 −16	**34**	63 −59	**35**	454 −293	**36**	817 −115

37 78+87   **38** 234+710   **39** 124+99   **40** 65+515

41 93−39   **42** 567−284   **43** 109−63   **44** 235−99

✓

Exercise 7B

1	385 55 +45	**2**	662 38 +102	**3**	34 613 +63	**4**	219 51 +134	**5**	222 444 +55	**6**	555 323 +44

7	75 102 128 +184	**8**	209 180 242 +243	**9**	221 344 357 +1728	**10**	1080 2916 2106 +1584	**11**	444 334 434 +222	**12**	666 555 444 +333

13	3172 −2765	**14**	4060 −3875	**15**	32 100 −30 407	**16**	67 571 −59 704	**17**	4170 −2264	**18**	5800 −3876

19	6231 −4268	**20**	5143 −2847	**21**	3421 −2426	**22**	5432 −2345	**23**	6543 −4544	**24**	1000 −888

25 79 + 21 + 84

26 352 + 178 − 177

27 420 − 75 + 123

28 503 − 49 + 8

29 305 − 94 + 167

30 536 + 521 − 9

BRAINSTRETCHER

1

Seve Ballesteros wins the British Open Golf Championship!

THE OLD COURSE
ST ANDREWS

SCORE CARD

TOTAL

HOLE	LENGTH	PAR	SCORE	HOLE	LENGTH	PAR	SCORE
1	370	4	4	10	318	4	4
2	411	4	4	11	172	3	4
3	352	4	4	12	316	4	4
4	419	4	4	13	398	4	4
5	514	5	4	14	523	5	4
6	374	4	4	15	401	4	4
7	359	4	4	16	351	4	4
8	178	3	2	17	461	4	4
9	307	4	4	18	354	4	3

a Calculate his score for the final round.
b How many strokes was his total score below the total par score for the round?
c How many birdies had he? (A score one under par for a hole is called a birdie.)
d How many bogeys had he? (A score one over par for a hole is called a bogey.)

e The lengths of the holes are in yards. What are the lengths of the longest and shortest holes?

f Copy and complete this table:

Length of hole (yards)	100–199	200–299	. . . .
Number of holes			

2 Rule   Replace * by + or −.

Example 2 * 1 can mean 2 + 1 = 3, or 2 − 1 = 1.

a Find the replacements for * that give the required answers.

(i) 3 * 2 * 1 = 6   (ii) 3 * 2 * 1 = 4   (iii) 3 * 2 * 1 = 2

There is a whole number other than 2, 4 or 6 which could be the answer to 3 * 2 * 1. What is it?

b Find replacements for *:

(i) 4 * 3 * 2 * 1 = 2   (ii) 4 * 3 * 2 * 1 = 4   (iii) 4 * 3 * 2 * 1 = 6

(iv) 4 * 3 * 2 * 1 = 8   (v) 4 * 3 * 2 * 1 = 10

Can you find a second solution for 4 * 3 * 2 * 1 = 4?

c Try to find two different ways of doing each of these:

(i) 5 * 4 * 3 * 2 * 1 = 3   (ii) 5 * 4 * 3 * 2 * 1 = 5

(iii) 5 * 4 * 3 * 2 * 1 = 7   (iv) 5 * 4 * 3 * 2 * 1 = 9

d Is it possible to find the same answer to 6 * 5 * 4 * 3 * 2 * 1 in three different ways?

CHECK-UP ON **NUMBERS IN ACTION**—1

1A   Adding and subtracting whole numbers

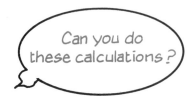

a 34 + 189

b 152 − 67

c £143 − £76 + £108

d Add two thousand five hundred and six to nine hundred and eighty-seven.

2A, B Problems with whole numbers

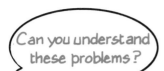

Can you understand these problems?

a How long are these routes?
  (i) A to B to D (ii) A to C to D
  (iii) A to B to C to D
  (iv) A to C to B to D

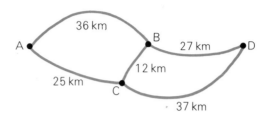

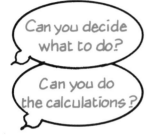

Can you decide what to do?

Can you do the calculations?

b A school prints 1800 programmes for a show. 1327 are sold, and 286 are given away. How many programmes are left?

c (i)   235   (ii)   1*7   (iii)   72   (iv)   258
     + ***   + 4*   −**   −1**
     ―――――   ―――――   ―――――   ―――――
      1000   *83   45   *90

3A, B, C Number sequences and patterns

Can you make two more patterns?

a

b

Can you find two more numbers in these? Can you give a rule for each?

c 91, 83, 75, 67, . . .

d 1, 4, 9, 16, . . .

e 4, 8, 6, 10, 8, 12, 10, . . .

Can you complete this magic square?

f

		6
	5	
4		

35

STATISTICS AND GRAPHS

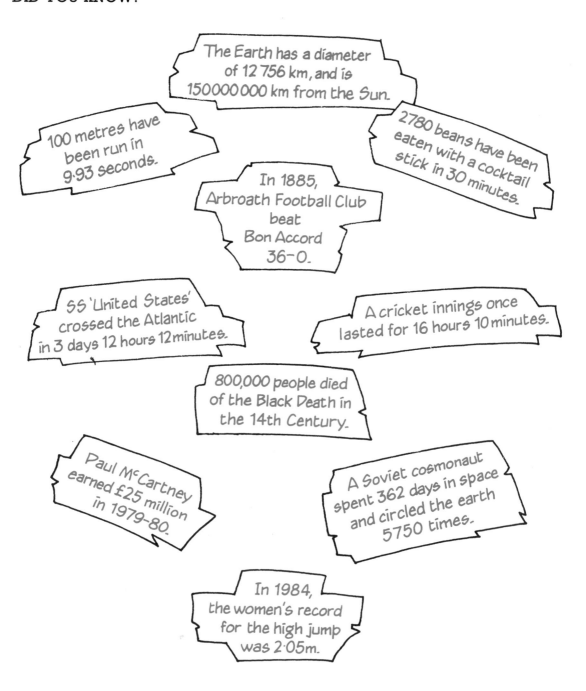

The Earth has a diameter of 12 756 km, and is 150 000 000 km from the Sun.

100 metres have been run in 9·93 seconds.

2780 beans have been eaten with a cocktail stick in 30 minutes.

In 1885, Arbroath Football Club beat Bon Accord 36-0.

SS 'United States' crossed the Atlantic in 3 days 12 hours 12 minutes.

A cricket innings once lasted for 16 hours 10 minutes.

800,000 people died of the Black Death in the 14th Century.

Paul McCartney earned £25 million in 1979-80.

A Soviet cosmonaut spent 362 days in space and circled the earth 5750 times.

In 1984, the women's record for the high jump was 2·05m.

Each of these numbers is called a **statistic**.
In this chapter you'll be gathering statistics, making tables of statistics and drawing diagrams of statistics taken from your school and from the world around you.

PICTOGRAPHS

In Alison's class 12 pupils walk to school, 8 cycle and 10 travel by bus or car. She drew a **pictograph**. This is a picture which shows these statistics more clearly.

Transport to school

Walk	☺	☺	☺	☺	☺	☺	☺	☺	☺	☺	☺	☺
Cycle	☺	☺	☺	☺	☺	☺	☺	☺				
Bus or car	☺	☺	☺	☺	☺	☺	☺	☺	☺	☺		

☺ = 1 pupil. This gives the *scale* of the pictograph

====== *Exercise 1 Statistics in the class* ======

1 Alison showed her pictograph to Petra. Petra liked it, and drew this one for the number of girls and boys in her class.
  a How many girls are in the class?
  b How many boys are in the class?
  c How many pupils are there altogether?

Girls and boys in class

Girls	🧍	🧍	🧍	🧍	🧍	🧍	🧍	🧍	🧍	🧍	🧍	🧍
Boys	🧍	🧍	🧍	🧍	🧍	🧍	🧍	🧍				

🧍 = 1 girl

🧍 = 1 boy

2 Petra's friend Martyn decided to draw a pictograph of the number of pupils in his class. There are 11 girls and 10 boys in the class. Draw his pictograph. Remember to give it a title and to show the scale, as in question **1**.

3 Count the number of girls and boys in your own class, and draw your own pictograph for this. Give it a title, and show the scale.

4 Soroya went round the class asking the pupils how they came to school each morning. She put the information in a table:

Walk	Bus or Car	Cycle
10	8	6

Draw a pictograph for her class.

5 Looking round the class, Muriel counted the number of pupils with different colours of hair. She drew a pictograph like this:

Hair colours

Brown	⛻	⛻	⛻	⛻	⛻	⛻	⛻	⛻	⛻	⛻	⛻
Black	⛻	⛻	⛻	⛻	⛻						
Fair	⛻	⛻	⛻	⛻	⛻	⛻	⛻	⛻	⛻	⛻	
Red/auburn	⛻	⛻	⛻								

⛻ = 1 pupil

a How many pupils are there in the class?
b How many have black hair?
c What is the most common colour?

6 Chris listened to pupils talking about their favourite subjects. Some said 'I like English best', or 'I like games', or another subject. So he asked all the pupils in his class for their favourite subjects. This is what he found.

English	Maths	French	Games	Art	Science
6	4	3	7	2	5

Copy and complete this pictograph he drew:

Favourite subjects

English	⛻	⛻	⛻
Maths	⛻	⛻	⛻
French	⛻	⛻	⛻
Games	⛻	⛻	
Art	⛻	⛻	
Science	⛻	⛻	

Exercise 2A Statistics around the school

1 Jeremy made a survey of the number of pupils who were members of school clubs. Here is his pictograph:

School clubs

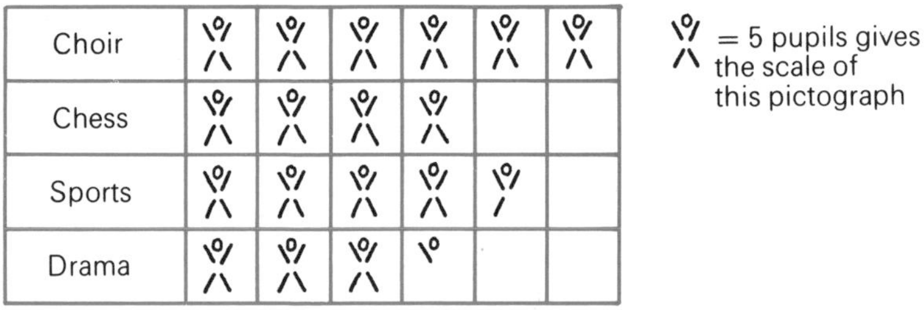

Choir	☆	☆	☆	☆	☆	☆
Chess	☆	☆	☆	☆		
Sports	☆	☆	☆	☆	☆	
Drama	☆	☆	☆	☆		

☆ = 5 pupils gives the scale of this pictograph

a Why do you think he made each figure stand for 5 pupils?
b How are 4 pupils shown in the table?
c How many pupils are there in each club?

2 Mr Able was asked by the headteacher to find out how many 12-year-old pupils took school lunches or packed lunches and how many went to a café or went home for lunch. He put his results in a table.

School lunch	Packed lunch	Café	Home
40	27	24	31

Draw a pictograph, taking  = 5 pupils.

3 The school shop sells soft drinks. The sales during one week are shown below.

Soft drinks sales

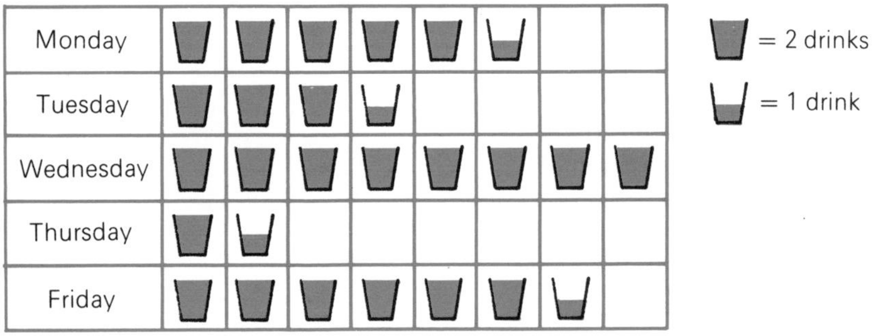

Monday	�U	�U	�U	�U	�U	u		
Tuesday	▼	▼	▼	u				
Wednesday	▼	▼	▼	▼	▼	▼	▼	▼
Thursday	▼	u						
Friday	▼	▼	▼	▼	▼	▼	u	

▼ = 2 drinks
u = 1 drink

a How many drinks were sold each day?
b A drink costs 18 pence. Calculate:
  (i) the highest day's takings   (ii) the lowest day's takings.
c How much money was taken during the week from sales of soft drinks?

STATISTICS AND GRAPHS

Exercise 2B   Statistics out of school

1 After the holidays 100 pupils were asked where they had been during the summer. This table shows the results.

At home	Abroad	England	Wales	Ireland	Scotland
15	25	18	8	10	24

Draw a pictograph, taking ⚊ = 5 pupils.

2 Many of the pupils go along to support their local football team, Action United. At the end of one season this pictograph was in the match programme to show the attendances at home games.

Number of spectators (to nearest hundred)

August	⚊	⚊	⚊	⚊				
September	⚊	⚊	⚊	⚊	⚊	○		
October	⚊	⚊	⚊	⚊	⚊	⚊	⚊	
November	⚊	⚊	⚊	⚊	⚊	⚊	⚊	⚊
December	⚊	⚊	⚊	⚊	⚊	⚊	⚊	
January	⚊	⚊	⚊	⚊	⚊	⚊	⚊	
February	⚊	⚊	⚊	⚊	⚊	⚊	○	
March	⚊	⚊	⚊	⚊	○			

⚊ = 500 spectators

a Why do you think the attendance was low in August?
b In which month was the attendance highest?
How many spectators were there that month?
c How many came to watch United in February?
d What was the total attendance for the season?
e How would you show attendances of 1759 and 2096?

3 One day Lin's class was working on a project. They had to take a note of the colours of the cars passing the school during half an hour.
20 were blue, 12 green, 25 red, 8 yellow, 5 black and 4 white.
Show these statistics in a pictograph, arranging it in order from the most common to the least common colour. Take 🚗 = 5 cars.

How many cars passed during this time?

Working on your own, or with a partner, or in a group, investigate some of the following for the pupils in your class.
Collect the statistics in a table, and then illustrate them with pictographs.

1 The ways in which they travel to school.

2 Their favourite sports.

3 Their favourite (or least-liked) subjects.

4 The months in which they were born.

BAR GRAPHS

Mr Sharp is still in charge of the school shop. He tries to order the things the pupils like best from the 'Fair Deal' Cash and Carry. One month he kept a record of the sales of crisps, soft drinks, lollipops, chocolate biscuits and sweets.
First of all, crisps:

Flavour	Cheese and Onion	Beef	Salt and Vinegar	Chicken	Plain	Prawn
Number of boxes sold	5	7	4	6	9	2

He then drew a bar graph.
He thinks that bar graphs
are easy to draw and to read.
Notice the titles on both axes,
and the scale on the vertical
axis.

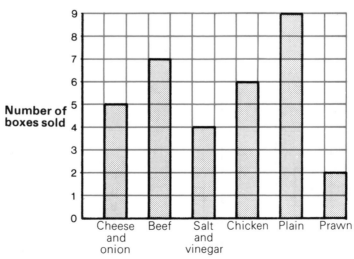

Exercise 3A   Statistics in the school shop

1 The next month the sales of crisps in the school were:

Flavour	Cheese and Onion	Beef	Salt and Vinegar	Chicken	Plain	Prawn
Number of boxes sold	6	8	3	5	10	1

  a On squared paper draw a bar graph like the one above. Make sure that you show the titles, and the scale at the side.
  b Which was the most popular flavour during these two months?
  c Which flavour would Mr Sharp probably not order again?

2 The record of sales of soft drinks during the first month was:

Flavour	Cola	Soda	Lime	Orange	Lemon
Number of crates sold	5	7	5	9	3

  a Using squared paper, copy and complete the bar graph.
  b Which type of drink had the highest sales?
  c List the flavours in order, from most popular to least popular.

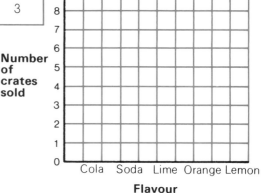

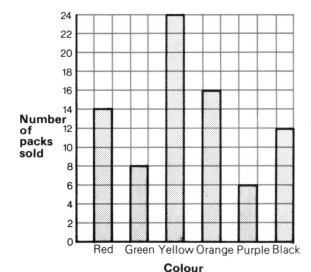

3 Lollipops of different colours sold well at lunch-time.
This bar graph shows the sales of lollipops during the month.
  a What colour was most popular?
  b What colour was least popular?
  c How many packs of yellow lollipops were sold?
  d Calculate the total number of packs of lollipops sold.

4 List all the things you can find wrong in these two graphs.

a

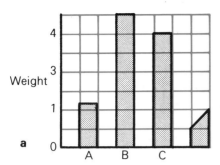

b

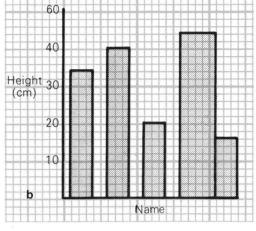

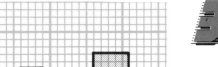

5 The pupils liked the boxes of sweets that were on sale in the school shop.
Here is the sales bar chart for sweets sold during the month.
  a Which sweet was the best seller?
  b How many boxes of Chews were sold?
  c How many more boxes of Mintos than Toffees were sold?
  d Calculate the total number of boxes of sweets sold.

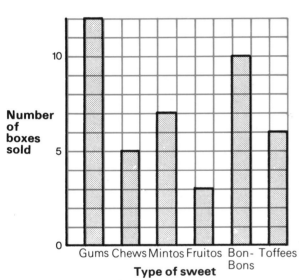

6 The school dentist didn't like it, but chocolate biscuits were also popular. Sales during the month were:

SNAP—4 dozen, YUMMY—1 dozen, FAB—7 dozen, CHEWY—10 dozen,

CRUNCH—8 dozen, FUDGE—3 dozen, NUTTY—5 dozen.

Show these sales in a bar graph.

7 Make a table showing the number of people in the families of the pupils in your class. Draw a bar graph to illustrate the data. Write a sentence about the graph.

STATISTICS AND GRAPHS

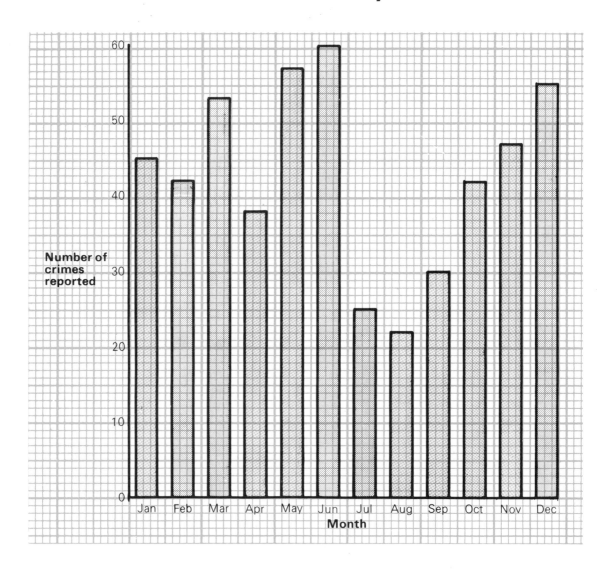

1 This bar graph shows the number of crimes reported in Hopetown last year.

a During which month were most crimes reported? How many were reported?
b In which two months were the same number of crimes reported?
c How many crimes were reported during the first six months of the year?
d How many were reported during the second six months?
e During the year extra police were sent to the town. In which month do you think they took up duty?

2 The manager of Hopetown's Home Furnishing Store had a meeting with his staff. He showed them this sales bar graph for the year.

a Which department was he pleased with?

b Which department was he not happy with?

c The manager said that the total sales were £40 000 more than the previous year's. What were the previous year's sales?

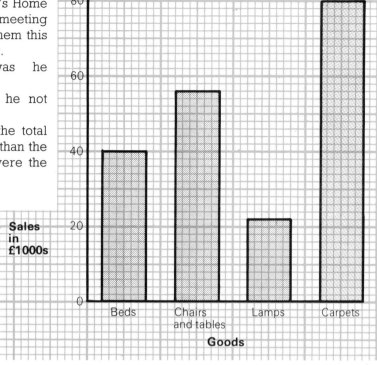

3 What vertical scales would you choose for data which go up to:

a 43 cm   **b** 100 g   **c** 1000 mm   **d** 2 kg?

4 This table gives the sunshine (to the nearest hour) and the rainfall (to the nearest mm) at Hopetown for each month during one year.

Month	Jan	Feb	Mar	Apr	May	June	July	Aug	Sep	Oct	Nov	Dec
Sunshine	60	71	92	135	187	172	159	166	158	135	99	75
Rainfall	135	129	110	105	83	70	74	65	51	70	105	120

a Choose suitable scales—do they have to start at zero?

b Draw bar graphs to illustrate the sunshine and rainfall records.

c Describe your graphs in one or two sentences.

5 Hopetown's Hi-Fi TV Superstore kept a record of the number of television sets sold or rented year by year.

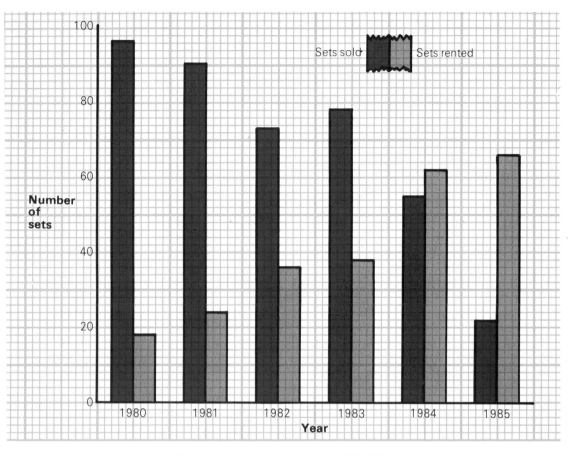

a How many sets were sold, and how many were rented, in 1985?

b In which year was the greatest drop in sets sold from the previous year? What was this drop?

c In which year was the greatest rise in sets rented from the previous year? What was this increase?

d In which year was the total number of sets sold and rented highest? What was this total?

e Write a sentence about the trends of the number of sets sold and rented.

===== *Exercise 3C   Statistics around the world* =====

Andrew found a book in the library with lots of statistics in it. Here are some of them.

1 Many different languages are spoken in the world. The numbers of people, to the nearest million, who speak some of them are shown below. Draw a bar graph to illustrate these statistics.

Mandarin	500 000 000	English	350 000 000	Hindi	145 000 000
Russian	130 000 000	Spanish	125 000 000	German	120 000 000
Japanese	116 000 000	Arabic	100 000 000	Bengali	100 000 000

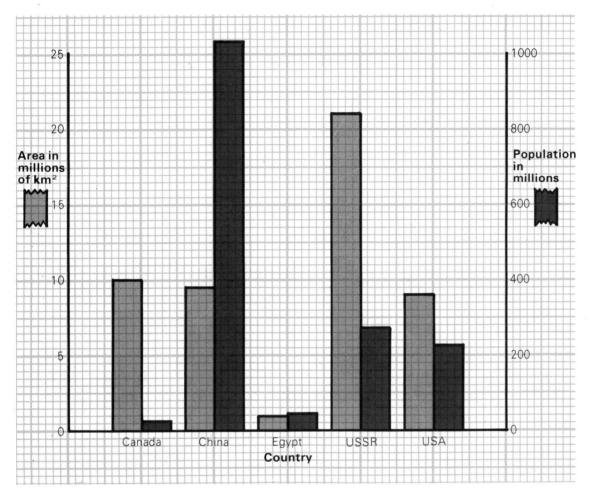

2 a Write down the areas of these countries in order, from largest to smallest.
  b Write down the sizes of their populations in order, from largest to smallest.
  c What do the pairs of bars in the graph tell you about the density of population in the countries?

3 People following some of the world's religions are believed to number, to the nearest million:

Buddhism	256 000 000	Christianity	1 200 000 000	Confucianism	275 000 000
Hinduism	500 000 000	Islam	1 100 000 000	Judaism	14 000 000
Shinto	98 000 000	Sikhism	14 000 000	Taoism	30 000 000

Draw a bar graph of these data.

GATHERING INFORMATION

=============== *Exercise 4A   More class statistics* ===============

1 Nick and Simon came into the maths classroom arguing about the popularity of the animals they kept as pets.

Mrs Page, their teacher, joined in. 'You'll need to collect some statistics', she said. They decided to do this for their own class by finding out which pets the pupils kept. Here are the results:

Pet	Tally	Number
Dog	ⅲ‖ ‖	7
Cat	ⅲ‖ ‖‖‖	
Budgie	‖‖‖	
Snake	‖	
Tortoise	‖‖	
Fish	ⅲ‖	
No pet	‖	

Look at the tally method of counting.
The 'barred gates' help us to count the totals quickly in 5s.
Copy the table, and fill in the last column.

Nick was quick to say 'I told you so'.
• Mrs Page said 'Don't be too sure. Try another class'.

2 Here are the statistics for the other class.

Pet	Dog	Cat	Budgie	Tortoise	Guinea pig	No pet
Tally	ⅲ‖ ⅲ‖	ⅲ‖ ‖‖‖	ⅲ‖	‖	‖	‖‖‖
Number						

Copy the table and fill in the bottom row.
Who said, 'I told you so', this time?

3 Gather the same kind of information for your own class. Put it in a table, using tally marks. Then draw a bar graph of the results.
What did you find out?

4

This is what they found.

	Tally	Number
Right-handed	⟋⟋⟋⟋ ⟋⟋⟋⟋ ⟋⟋⟋⟋ ⟋⟋⟋⟋ ⟋⟋⟋⟋	
Left-handed	⟋⟋⟋	

a Copy the table and fill in the last column.
b Do the figures agree with what Mrs Page said?
c How many pupils are in the class?

5 Here are the colours of the pupils' eyes in Anne's class:
Blue, blue, brown, grey, blue, green, grey, blue, brown, hazel, grey, grey, blue, grey, blue, brown, grey, green, hazel, grey, blue, blue, blue, blue, grey, brown, blue.
Copy and complete this tally table:

Colour	Blue	Green	Grey	Brown	Hazel
Tally					
Number					

6 Apart from class statistics, can you think of other statistics that would be worth gathering and showing in graphs?

Exercise 4B School and sports statistics

1 By this time Mrs Page's class had learned its lesson. So when Billy said, 'Most people walk to school', they all knew what to do.
Here are their replies:
Walk, walk, cycle, bus, walk, bus, bus, cycle, bus, car, walk,
walk, cycle, bus, bus, bus, walk, bus, cycle, car, walk, car,
cycle, walk, walk, walk, bus, walk, car, walk.
Make a table like this, and fill it in:

	Walk	Cycle	Bus	Car
Tally				
Number				

Do your results agree with what Billy said?

STATISTICS AND GRAPHS

2 Mr Davis taught two classes of 12-year-olds. He thought that class A1 was better at French than class A2.

Every Friday he gave them a test, marked out of 10.

a One Friday, the marks for A1 were:

1, 5, 0, 4, 3, 5, 1, 6, 5, 4, 4, 4, 4, 5, 4, 2,
4, 5, 3, 4, 9, 2, 2, 3, 4, 0, 7, 3, 1, 4, 10, 4

Mark	0	1	2		
Tally					
Number					

Make a tally table, and complete it.

b On the same day the marks for A2 in the same test were:

10, 9, 8, 7, 0, 1, 3, 4, 5, 4, 8, 9, 10, 1, 2, 5,
4, 5, 5, 5, 7, 3, 2, 1, 8, 8, 9, 6, 4, 4, 6, 7

Make a table like the one above.

c Draw a marks bar graph for each class, with MARKS on the horizontal axis and NUMBER on the vertical axis.

d Do you think Mr Davis was correct? Give a reason for your answer.

3 Jean was captain of the Junior Hockey Team. Eighteen games were played during the season. Jean kept a note of the scores. She always wrote the goals scored by her team first:

2–1, 1–1, 2–1, 4–2, 0–0, 0–2, 3–4, 2–2, 5–3,
1–0, 4–0, 2–4, 1–1, 1–2, 0–3, 3–1, 2–1, 0–2

a Copy and complete Jean's table.
b How many points did her team gain if there were 2 points for a win and 1 for a draw?
c Do you think her team had a good season? Explain your answer.

	Tally	Number of games
Won		
Drawn		
Lost		
		18

4 Jean wondered whether her team was stronger in attack than in defence.
a Add up the number of goals scored by the team, and the number scored against them.
b What do you think?

5 Paul went on holiday for 10 days. Each day he had a round of putting of 18 holes. Here are two of his score cards.

First day

HOLE	SCORE	HOLE	SCORE	
1	2	10	3	
2	3	11	4	
3	1	12	2	
4	4	13	4	
5	3	14	3	
6	4	15	2	
7	3	16	4	
8	3	17	2	
9	2	18	5	TOTAL
TOTAL		TOTAL		

Last day

HOLE	SCORE	HOLE	SCORE	
1	2	10	2	
2	3	11	3	
3	2	12	1	
4	4	13	5	
5	1	14	3	
6	3	15	2	
7	4	16	4	
8	3	17	2	
9	2	18	4	TOTAL
TOTAL		TOTAL		

a Calculate his total score for 18 holes on each of these two days.
b Calculate his average score per hole in each game.
c Had he improved his game during his holiday?

6 Paul wanted to see the number of 1s, 2s, . . . he scored. So for *each* day he made a table like this:

Score	1	2	
Tally			
Number of times			

a Make up the two tables, and draw a bar graph for each one.
b Do you think these show that Paul's game had improved?

---

a Make a table like this, with all the letters of the alphabet listed. Which do you think are the most common and the least common letters?

Letter	Tally	Number
a		
b		
c		

b Choose a page in a book, and enter a tally mark in your table opposite each letter as it occurs.
c Complete the table, and write a sentence about the letters that appear most often and least often on the page.
d Compare your results with those of other pupils in your class.

STATISTICS AND GRAPHS

1A   Pictographs

Can you read and draw pictographs?

This pictograph shows the number of pupils absent from school each day in one week:

Monday	👤	👤	👤	👤	👤	👤	9	
Tuesday	👤	👤	👤	👤	👤	9		
Wednesday	👤	👤	👤	9				
Thursday	👤	👤	👤	°				
Friday	👤	👤	👤	👤	👤	👤	👤	👤	9

👤 = 5 absent pupils

a On which day were there fewest absences? How many were there?
b On which day were there most absences? How many were there?
c How many absences were there altogether that week?

2A, B   Bar graphs

Can you read and draw bar graphs?

The label on a box of SPARKY matches says there are 45–47 matches in each box.
The number of matches in each of 50 boxes was counted. Here are the results.

Number of matches	42	43	44	45	46	47	48
Number of boxes	4	7	5	10	12	9	3

a Draw a bar graph of the results.
b How many boxes did not contain 45–47 matches?

3A, B   Gathering information

Can you arrange information in a table?

At a dog show, owners were asked which dog food they usually bought. The replies were:
GO, PET, PET, GO, FAB, GO, WAG, PAWS, WAG, PET, GO, BARK, WAG, GO, PAWS, BARK, PET, GO, BARK, PET, PAWS, PAWS, GO, GO, WAG, PET, PAWS, PET, GO, PET.
a Make a table, using tally marks.
b Draw a bar graph of the results.
c Which is the most popular dog food?

X MARKS THE SPOT

Class discussion

1 There is treasure buried on this island at the place marked X.

Imagine that some people have just made a safe landing on the island.

What directions would you give them to help them find the treasure?

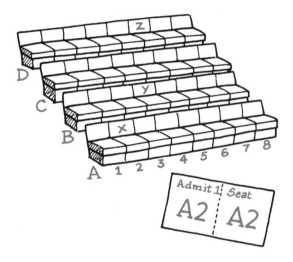

3 a Using the rows and columns, say where Ian sits in this classroom.
  b James sits beside Ian. Using rows and columns, say where James sits.
  c In the same way, say where Meena sits.
  d Who sits in row 4, column 1?
  e Who sits in row 1, column 1?
  f In the same way, say where you sit in your class.

2 Here are the front rows of seats in a theatre. John's ticket for seat X is shown. Lesley and Donna have tickets for seats Y and Z.

What would their tickets have on them?

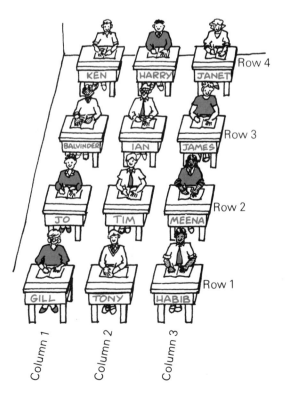

53

COORDINATES

4 Town maps often have squares drawn on them. This helps you to find places in the town. On this map the Bus Station is in square (E, 7).

In which squares would you find these places?
a The Castle
b Holyrood Palace
c The Railway Station
d The Museum.
Which named roads meet in these squares?
e (J, 10)
f (A, 5)
If you were on your way to (G, 5) what might you take with you?

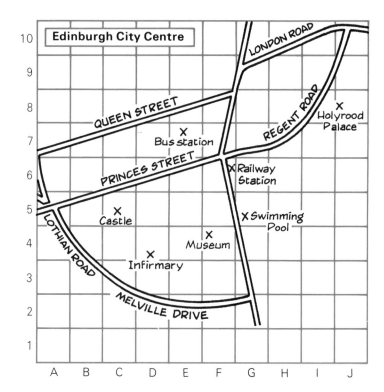

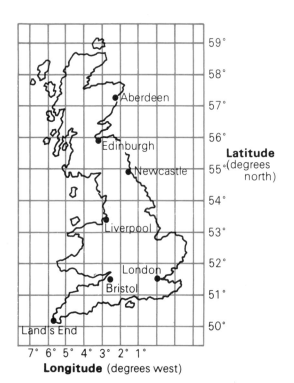

5 To find any place on earth on a map you need to know its latitude and longitude.
This map of Britain is marked off in rectangles. The bottom left corner fixes the position of the rectangle.
So Liverpool is in the rectangle (3°W, 53°N). Which rectangles are these in?
a Edinburgh
b Bristol
c Aberdeen
d Newcastle
e Land's End.
The line through London has a longitude of 0°.
This line is known as the Greenwich Meridian.
Do you know the name of the 'line' which has a latitude of 0°?
Do you know the name of the place which has a latitude of 90°N?

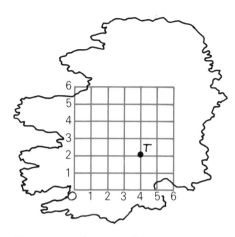

The coordinate grid

Instead of east, we have the X-direction.
Instead of north, we have the Y-direction.
OX is the x-axis, and OY is the y-axis.
O(0, 0) is the origin.
The point T (4, 2) has x-coordinate 4 and
y-coordinate 2.

In the 17th century a Frenchman called René
Descartes used the same idea to fix the
position of a point.
Treasure Island in the first question could
be marked off in squares like this. O is the
safe landing.
The treasure T is 4 km east and 2 km north
of O.
Descartes would have said that T was the
point (4, 2).

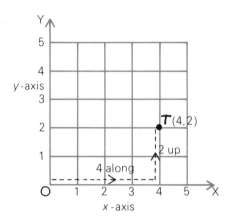

Exercise 1A Treasure hunts

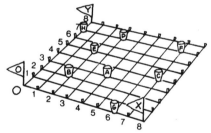

A treasure hunt at a fête is laid out using
string and pegs. You have to guess the
crossing where the treasure is hidden.

Using x- and y-axes it looks like this.

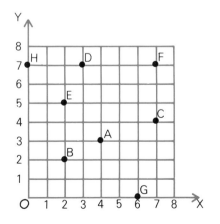

1 Identify these guesses. For example,
A(4, 3). (The x-coordinate is always given
first.)

  a Adam guesses that the treasure is at the
  point A(4, . . .).
  b Bill's guess is B(. . ., 2).
  c Catriona's guess is C(7, . . .).
  d Dave's guess is the point D with coordinates (. . ., . . .).
  e The x-coordinate of Erica's guess is . . .
  and the y-coordinate is . . . .
  f Fatima's guess is (. . ., . . .).
  g The y-coordinate of Gordon's guess
  is . . . .
  h Harry's guess has an x-coordinate of . . . .

2 In a new game, eight people try to guess where the treasure is. Copy and complete the list of coordinates of their guesses.
E(8, . . .), F(1, . . .), G(5, . . .),
H(. . . , 6), J(0, . . .), K(. . . , . . .),
L(. . . , . . .), M(. . . , . . .).

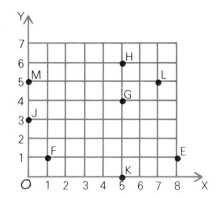

3 Prizes are buried at the points S(3, 3), T(4, 0), U(0, 6), V(5, 1) and W(1, 4).
On squared paper, draw axes OX and OY at right angles, and number them from 0 to 6.
Plot the positions of the prizes.

4 A sign-writer uses coordinates to design his letters.
Plot each of these sets of points.
Join them up as shown by the arrows to find the letters.
  a $(4, 9) \rightarrow (5, 6) \rightarrow (6, 9)$
  b $(1, 9) \rightarrow (1, 6) \rightarrow (3, 6)$
  c $(0, 2) \rightarrow (0, 5) \rightarrow (2, 2) \rightarrow (2, 5)$
  d $(6, 6) \rightarrow (8, 6) \rightarrow (6, 3) \rightarrow (8, 3)$
  e $(3, 0) \rightarrow (3, 3) \rightarrow (4, 1) \rightarrow (5, 3) \rightarrow (5, 0)$

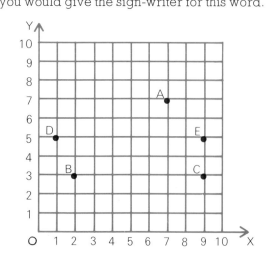

5 On squared paper draw axes OX and OY.
Number OX from 0 to 10, and OY from 0 to 4.
Plot these points and join them up in the order in which they are listed. If you add an eye and a mouth you'll have a . . . . (0, 3), (7, 3), (8, 4), (9, 4), (9, 3), (10, 3), (9, 2), (8, 2), (8, 1) (7, 1), (7, 0), (6, 1), (3, 1), (2, 0), (2, 2), (0, 3).

6 On squared paper draw the word EXIT.
Write down the coordinates of all the points you would give the sign-writer for this word.

7 New roads are being planned which will pass through a country area.
A map is drawn to show the plans for the new roads.

A is an abandoned cottage.
B is a barn.
C is a ruined castle.
D is a farmhouse.
E is a croft.

a On squared paper draw axes OX and OY, and number them from 0 to 10. Plot the five points A to E.

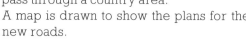

b The first road will pass through the points O(0,0), F(2,2), G(4,4), H(6,6), I(8,8) and J(10,10).
Plot these points and draw a straight line through them.

c Is there anything in the road's planned route?

d The second road will pass through the points P(10,0), Q(8,2), R(6,4), S(4,6), T(2,8) and U(0,10).
Plot these points and draw a straight line through them.

e Give the coordinates of the point where a crossroads sign will be needed.

Hit the target!

You need a dice for this game.
Try playing with your neighbour!
Throw the dice twice. The first score gives the *x*-coordinate, and the second the *y*-coordinate, of your shot.

 then → (4, 3).

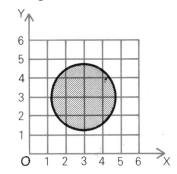

This would count as a hit on the target.
Take turns with your neighbour. The first person to score four hits is the winner.

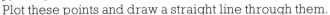

Exercise 1B Robots and dancers

1 In some factories, machines for cutting shapes are controlled by computer.
To give the machine its orders the computer must be told the coordinates of the corners of each shape.
Write down the name and coordinates of each corner of the shapes shown below like this, A(1, 2).

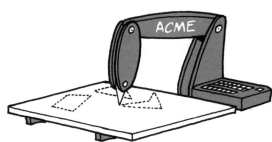

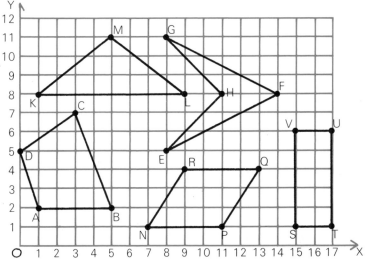

2 A logo was designed for the front of the programme for 'The Old-Time Variety Show'.

To draw this logo, plot these points on squared paper and join them up in the order given.

(4, 1), (12, 1), (8, 5), (8, 8), (6, 10), (6, 13), (8, 14), (7, 16), (6, 16), (4, 15), (5, 13), (1, 13), (0, 8), (2, 10), (1, 13), (3, 11), (3, 10), (4, 9), (4, 1).

When there are several points to list, as in question **2**, they can be put in a table, like this.

x	2	4	5
y	3	6	9

3 A robot arm bores holes for screws in a metal plate. It is programmed to bore holes in these positions:

x	1	8	8	1	2	7	2
y	1	1	7	7	2	4	6

Draw a set of axes to represent two edges of the plate. Plot the positions of the holes.

4 The same robot arm can cut curves in the metal if it is told at least five points that the curve passes through. Plot the points and join them up with a smooth curve.

x	0	1	2	3	4
y	0	1	4	9	16

5 Repeat question **4** for these points:

x	1	2	4	8	16
y	16	8	4	2	1

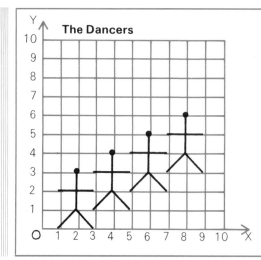

The Dancers

While designing wallpaper an artist draws a repeated pattern of 'stick' dancers.

Copy this picture of the stick dancers.

Look carefully at how one dancer follows another. Draw four more dancers in the pattern and write down the coordinates of the eight heads.

Look for a pattern in the coordinates. Write down the coordinates of the head of the twentieth dancer.

If a dancer's head is at (a, b), how are a and b connected? Investigate the sequence of positions of each dancer's feet.

SETS OF POINTS

Exercise 2A Pirates and plots

1 Pirate Pete is in a fix. Here he is under the palm tree at the crossing of Skullbones Path and Deadman's Ditch. Where is the treasure? He only has one half of the map. This one!

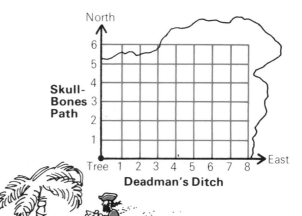

The Treasure is buried under the sand six paces East of Skullbones Path and

a Copy this diagram on squared paper.
b Plot four points where Pete might dig.
c Join these points, and continue the line in both directions.

Poor Pete: the treasure is somewhere on this line. But where?

Bluebeard Bill has the other half of the message.

five paces North of Deadman's Ditch

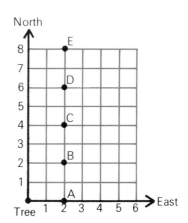

d Plot four points where Bill might dig.
e Join these points, and continue the line in both directions.

Only **you** know where the treasure is buried. Write down the coordinates of this point.

2 Mike the Mug was given a false message. He dug holes in the positions shown. Write down the coordinates of each point. What do you think Mike's message was?

3 On squared paper, draw x- and y-axes.
Plot the points $(5, 0)$, $(5, 1)$, $(5, 2)$, $(5, 3)$, $(5, 4)$, $(5, 5)$ and $(5, 6)$.
Draw a straight line through the points.

What do you notice about the x-coordinate of every point?

4 Plot the points $(0, 4)$, $(1, 4)$, $(2, 4)$, $(3, 4)$, $(4, 4)$ and $(5, 4)$ on the same diagram.
Draw a line through the points.
Make up a rule which fits every point on this line.

5

Rules: (i) The design must be based on this set of points:
$(0, 0)$, $(8, 0)$, $(8, 3)$, $(8, 6)$, $(0, 6)$, $(0, 3)$.
(ii) Any pairs of these points may be joined by straight lines.
(iii) The flag must be coloured.
Try designing one or two flags.

6 Copy this diagram on squared paper. Plot the points $(1, 1)$, $(2, 3)$, $(3, 2)$, $(5, 3)$, $(4, 4)$, $(3, 4)$ and $(0, 2)$.
Which of these points lie inside the shaded square?
Which points lie outside the square?
Copy and complete these rules:
a A point inside the square has an x-coordinate greater than . . . , but less than . . . .
b A point inside the square has a y-coordinate greater than . . . , but less than . . . .

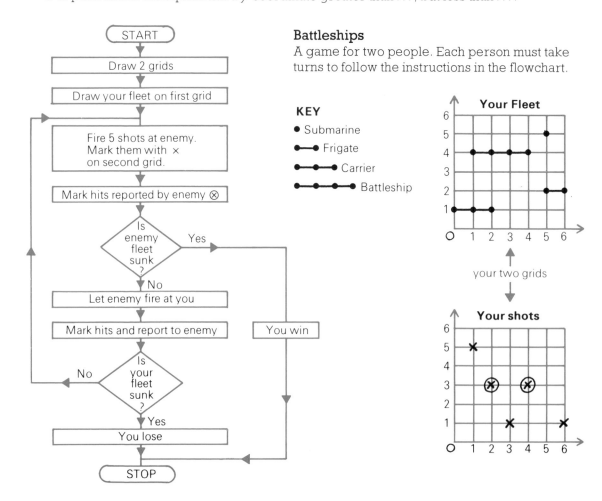

Battleships
A game for two people. Each person must take turns to follow the instructions in the flowchart.

KEY
● Submarine
●—● Frigate
●—●—● Carrier
●—●—●—● Battleship

1 Robbers based at R rob a bank at B.
The lines show a network of roads. Copy this on squared paper.
The police set up road-blocks at (1, 2), (2, 3), (3, 0), (3, 4), (4, 1), (4, 3) and (5, 2).
a Mark the roadblocks.
b Can the robbers escape?
c If you think the robbers can escape, draw their shortest route from B to R.
d At what other point should the police have set up a roadblock?

2 In the 'Hunt the Treasure' game in Exercise *1A* the treasure was at the point (3, 2). Mark this point on a diagram.
Now imagine that the point had an *x*-coordinate less than 3 and a *y*-coordinate less than 2.
Shade the area where it could have been.

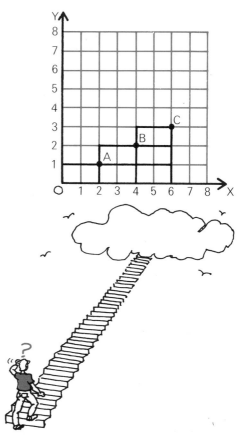

3 Here is a diagram of the start of a staircase. Copy it on squared paper and draw the next three stairs up.
a Write down the coordinates of A, B and C.
b Write down the coordinates of the next four points in the sequence.
c If the sequence continued, which of these points would belong to it?

   (40, 80), (20, 10), (700, 350), (200, 150).

d If (*p*, 40) is a member of the sequence, what is the value of *p*?
e If (16, *q*) is a member of the sequence, what is the value of *q*?
f If (*a*, *b*) belongs to the sequence, which of the following is true?

   (i) $b = 2 \times a$ (ii) $a = 2 \times b$ (iii) $a = \frac{1}{2}b$.

COORDINATES

4 Draw the line consisting of the set of points with *x*-coordinate 10.
Draw the line consisting of the set of points with *y*-coordinate 8.
Write down the coordinates of the point where the two lines meet.

5 a On squared paper draw the lines consisting of:

    (i) the set *A* of points with $x = 6$ (ii) the set *B* of points with $x = 9$

    (iii) the set *C* of points with $y = 4$ (iv) the set *D* of points with $y = 7$.

  b Write down the coordinates of the points common to the lines consisting of:

    (i) sets *A* and *C*   (ii) sets *A* and *D*   (iii) sets *B* and *C*   (iv) sets *D* and *B*.

  c Do the sets *A* and *B* have a point in common?
Explain your answer.

═══════════════════ *Exercise 2C   More or less* ═══════════════════

1 Plot the points $(0, 0)$, $(1, 1)$, $(2, 2)$, $(3, 3)$ and $(4, 4)$, and draw a straight line through the points.
P is the point (r, s). Find a rule which tells you if:
a P lies on the line
b P lies above the line
c P lies below the line.

2 Repeat question **1** for the points:
a $(0, 0)$, $(1, 2)$, $(2, 4)$ and $(3, 6)$
b $(0, 1)$, $(1, 5)$, $(2, 9)$ and $(3, 13)$.

3 *S* is the set of points with *x*-coordinate greater than 2 but less than 8.
T is the set of points with *y*-coordinate greater than 1 but less than 5.
To win a game you have to roll a coin so that it stops in the area where the sets *S* and *T* overlap.
Draw a diagram, and shade the winning area.

4 O is the origin, and OPQR is a different square in each of the following.
a P is the point $(3, 0)$. Write down the coordinates of Q and R.
b Q is the point $(5, 5)$. Write down the coordinates of P and R.
c R is the point $(0, k)$. Write down the coordinates of P and Q.
d If Q is the point (a, b), what can you say about *a* or *b* or both?
e If the perimeter of the square is greater than 36 units, and Q is the point (x, y), what can you say about *x* or *y* or both?

A computer program is needed for making a circle on the screen.

The instruction 'Plot (. . . , . . .)' is to be used, and a 40 × 20 graphics grid is available for plotting the points which will make up the circle.

Choose *x*- and *y*-axes on squared paper as two sides of the grid, and draw the largest possible circle on this grid.

Mark points on the circle, along with their coordinates, for use in the computer program.

Join up the points to see if they form a shape like a circle. If not, mark more points and their coordinates. How could you make the circle even more accurate?

Perhaps the best set of points could be used on one of the school's computers.

The National Grid

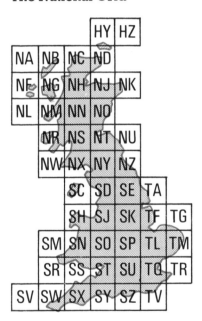

Maps often have the Ordnance Survey National Grid on them. Great Britain is divided into 100 km squares, named by pairs of letters—see diagram on the left.

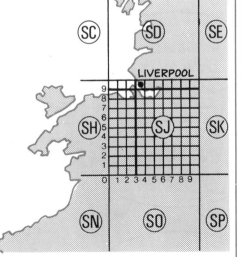

Each of these squares is divided into one hundred 10 km squares whose sides are numbered from 0 to 9—see diagram on the right.

A place can be found from the coordinates of the bottom left corner of the square it is in. For example, Liverpool—SJ39.

Investigate this system in an atlas or AA or RAC handbook. Describe it for your own area, and then use it to list the positions of towns of your choice.

COORDINATES

1A Copy this diagram, with all the dotted lines and boxes, into your notebook.

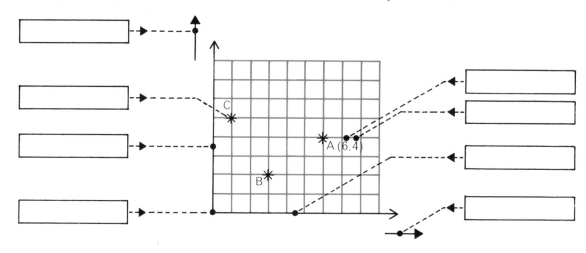

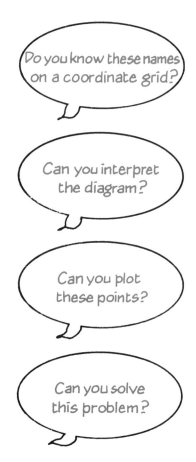

Do you know these names on a coordinate grid?

Can you interpret the diagram?

Can you plot these points?

Can you solve this problem?

2A, B a Fill in the boxes in your diagram, using these labels: *x*-direction, *x*-axis, *x*-coordinate of A, *y*-direction, *y*-axis, *y*-coordinate of A, the point C, the origin.

b Which of the points A, B or C has the greatest *x*-coordinate?

c Which of the three points has coordinates which add up to 6?

d A, B and C are the corners of a square. Write down the coordinates of the fourth corner.

3B, C a Draw axes OX and OY, and plot the points P(6, 2) and Q(2, 6). Draw the line PQ. Write down the coordinates of the point R at the middle of PQ.

b A line is drawn from O to R, and is then drawn the same distance to S. Write down the coordinates of S.

c Find the coordinates of three more points in the sequence R, S, . . . , and explain the method you used.

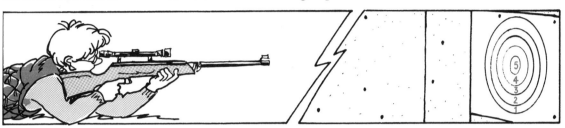

SPORT AND PASTIMES

Exercise 1A   Target practice

1 Three friends, Sam, Kerry and Gerald, go along to the Rifle Club. Each fires 10 shots at the target.

How many points did they each score?

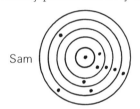

Sam

Kerry

Gerald

2 They were all disappointed with their scores and decided to try again.

Sam	**Kerry**	**Gerald**
2 hits in 1 point ring	1 hit in 1 point ring	2 hits in 1 point ring
1 hit in 2 points ring	2 hits in 2 points ring	0 hits in 2 points ring
3 hits in 3 points ring	3 hits in 3 points ring	4 hits in 3 points ring
2 hits in 4 points ring	2 hits in 4 points ring	2 hits in 4 points ring
2 hits in 5 points circle	2 hits in 5 points circle	1 hit in 5 points circle

a How many points did they each score this time? Arrange your work in a table.

b Who improved their scores?

c What happened to Gerald's score?

3 What is the highest score possible with 10 shots?

4 The three friends thought they were ready for a competition. Any entrant scoring more than 80 points with 25 shots would reach the final. Here are their scores. Who reached the final?

Number of hits, each worth—

	1 point	2 points	3 points	4 points	5 points
Sam	3	6	3	7	6
Kerry	6	0	8	4	5
Gerald	0	5	8	8	3

Exercise 1B Play golf

1 Karen, Bobby and Wendy played a round of golf. Their score-cards showed that:

Karen had 2 threes, 6 fours, 8 fives and 2 sixes.

Bobby had 3 fours, 6 fives, 3 sixes, 4 sevens and 2 nines.

Wendy had 2 fours, 7 fives, 4 sixes, 4 eights and 1 nine.

Calculate their scores for the round. Arrange your work in a table.

2 In need of practice, they tried the putting green. Calculate their scores from this table.

	Twos	**Threes**	**Fours**	**Fives**	**Sixes**
Karen	**7**	**8**	**3**		
Bobby	**2**	**8**	**7**		**1**
Wendy	**4**	**10**	**3**	**1**	

3 Bobby boasted that he would beat Karen and Wendy at 'Crazy Golf'. Did he manage to win this time?

Karen had 5 threes, 4 fives, 6 sixes and 3 eights.

Bobby had 14 fours, 3 fives and 1 nine.

Wendy had 3 threes, 8 fours, 5 sixes, 1 seven and 1 eight.

4 Bobby and Wendy enjoyed their golf so much that they decided to buy second-hand sets of golf clubs.

Bobby bought four woods at £12 each and seven irons at £8 each.

Wendy bought three woods at £13 each and six irons at £9 each.

How much did they pay for their sets of clubs?

Exercise 2A Bull's-eye

Look at the dart board. Why are two of the rings called double and treble?

1 Find the score if your dart lands in:

a Double 4   **b** Double 7   **c** Treble 7

d Treble 3   **e** Treble 8   **f** Double 15

g Treble 12   **h** Double 18   **i** Treble 14

j Double 19   **k** Treble 17   **l** Treble 19

What is the highest score possible with one dart?

2 Robert and Liz are playing darts. Robert needs 7 to win and Liz needs 9. Each has three darts, and the game must finish with a double.

a How could Robert win with his next two darts?

b Robert threw a 3 and a double 20. Liz went on to win the game. How could she do this?

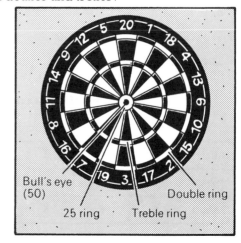

Bull's eye (50)

25 ring   Treble ring   Double ring

3

Jocky	Sandy	Keith	Fiona	Dave	Lola
3	Double 14	Treble 13	Double 6	Treble 9	Double 19
Double 20	9	Double 15	19	Double 14	13
5	Double 11	7	Double 17	18	Treble 8

Who plays first?

4 With three darts George scores 9, double 9 and treble 9.
Calculate George's total score in two different ways. Which way is quicker?

5 Pat scores treble 12, 12 and double 12 with three darts. Find his total in two ways.

Exercise 2B   Win a Fortune

The rules in the TV Quiz Game 'Win a Fortune' are:

Round 1 Each correct answer scores 10 points.

Round 2 Each correct answer is now worth the total number of points scored in Round 1. The contestants win £1 for each point of their final score.

Example Jayne answered 6 questions correctly in Round 1.
So she scored $6 \times 10 = 60$ points in Round 1.
She answered 5 questions correctly in Round 2.
So she scored $5 \times 60 = 300$ points.
She won £300.

1 Debbie answers five questions correctly in Round 1, and seven questions correctly in Round 2. How much does she win?

2 Five contestants in the Quiz Game answer the following number of questions correctly:

	Mary	John	Lucy	Peter	Simon
Round 1	6	8	9	6	7
Round 2	8	4	7	2	8

How much money does each win?

3 There are ten questions in each round of the Quiz Game.
If you gave two wrong answers in Round 1 and four wrong answers in Round 2, how much money would you win?

4 Five friends entered the quiz and scored as follows:

	Nigel	Salim	Keith	Laura	Penny
Round 1	4	7	8	5	9
Round 2	7	4	5	8	6

a How much money did each win?
b In what other way could you win £540 in the quiz?
c Write down two ways in which you could win £350.

5 Amanda was very excited. She answered all ten questions correctly in both rounds. How much had she won?

─────────────── **Exercise 2C What a driver!** ───────────────

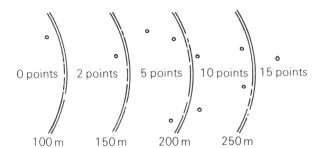

At a golf driving range players are given 50 golf-balls to hit as far as they can. The scores are shown in the drawing. For example, 100–150 metres (including 100 but not 150) scores 2 points.

1 Paul has driven the ten golf-balls shown above. What is his score so far?

2 Jack and Nicholas each have a turn on the driving range. Here are their results.

	Distance in metres			
	100–149	150–199	200–249	250 and over
Jack	16	22	5	1
Nicholas	14	19	12	3

a How many points did each boy score?
b How many non-scoring shots did each have?

3 Alastair played on his own and scored as follows:

Less than 100 m	100–149 m	150–199 m	200–249 m
1	16	23	7

a How many times did he drive a ball 250 metres or further?
b What was his score?

4 The highest score ever recorded at the driving range was 493. To reach this score the record-breaking player hit 14 shots of 250 or more metres, 23 shots between 200 and 249 metres, and 9 shots between 150 and 199 metres.

  a How many shots were hit a distance between 100 and 149 metres? Arrange your work in a table.

  b How many shots failed to travel 100 metres?

MAKING SURE OF MULTIPLICATION

Any number × 0 = 0
0 × any number = 0

(i) Multiplication by zero

Be careful when multiplying by zero!

Examples: $\quad 2 \times 0 = 0$
$\qquad\qquad 673 \times 0 = 0$
$\qquad\qquad 0 \times 1 = 0$

(ii) Multiplication by 10, 100, 1000

T	U
	6
6	0

×10

= (for second row)

H	T	U
	1	5
1	5	0

×10

Th	H	T	U
	2	4	7
2	4	7	0

×10

When a whole number is multiplied by 10, each figure moves 1 place to the left.
Units become tens, tens become hundreds and hundreds become thousands.

Th	H	T	U
		3	5
3	5	0	0

×100

2 places to left

Th	H	T	U
			7
7	0	0	0

×1000

3 places to left

Multiply by 10, 100, 1000.

Answer is a larger number.

HINT To multiply by 20, multiply by 2, then by 10.

Example: $57 \times 20 = 57 \times 2 \times 10 = 114 \times 10 = 1140.$

(iii) Products

In $4 \times 7 = 28$, 28 is the product of 4 and 7.
Sometimes there are easy ways to calculate products.

Example: $2 \times 39 \times 50 = 39 \times (2 \times 50) = 39 \times 100 = 3900.$

Calculator out of action again! Battery done. Do Exercises *3A* and *3B* mentally, or using paper and pencil.

======================== *Exercise 3A* ========================

1 Multiply by zero:
   a 9   **b** 14   **c** 87   **d** 129   **e** 1000

2 Multiply by 10:
   a 1   **b** 17   **c** 66   **d** 101   **e** 5321

3 Multiply by 100:
   a 8   **b** 80   **c** 234   **d** 1   **e** 0

4 Multiply by 1000:
   a 23   **b** 1   **c** 0   **d** 100   **e** 1000

5 Calculate:
   a 94×10 **b** 38×100 **c** 64×0 **d** 85×1000 **e** 10×100

   f 50×100 **g** 10×204 **h** 91×1000 **i** 100×100 **j** $24\,680 \times 0$

6 Calculate:
   a 6×5 **b** 6×7 **c** 8×8 **d** 5×9 **e** 7×7

   f 7×9 **g** 5×8 **h** 8×4 **i** 9×3 **j** 9×9

   k 54×3 **l** 54×30 **m** 54×300 **n** 54×3000 **o** 54×0

======================== *Exercise 3B* ========================

1 Calculate:
   a 100×100 **b** 1000×0 **c** 240×10 **d** 100×204 **e** 0×2468

2 Calculate:
   a 45×2 **b** 27×3 **c** 33×4 **d** 84×5 **e** 64×6

   f 71×7 **g** 35×8 **h** 32×9 **i** 99×0 **j** 123×5

   k 4×47 **l** 40×47 **m** 400×47 **n** 4000×47 **o** 0×47

3 Find these products:
   a 17×2 **b** 17×20 **c** 17×200 **d** 17×0 **e** 0×17

   f 5×62 **g** 4×74 **h** 6×76 **i** 62×8 **j** 599×1

   k 82×20 **l** 76×50 **m** 54×60 **n** 33×70 **o** 132×80

4 Calculate:
   a $2 \times 7 \times 5$ **b** $4 \times 6 \times 10$ **c** $5 \times 19 \times 2$ **d** $2 \times 17 \times 10$

   e $4 \times 25 \times 100$ **f** $25 \times 16 \times 4$ **g** $19 \times 3 \times 10$ **h** $5 \times 17 \times 20$

   i $2 \times 33 \times 50$ **j** $5 \times 20 \times 0$ **k** $50 \times 41 \times 2$ **l** $0 \times 111 \times 0$

MAKING USE OF MULTIPLICATION

=== *Exercise 4* ===

1 Ali's new bike will cost him £3 a week for
35 weeks.
How much is this altogether?

2 A school has nine classes of twelve-year-old pupils. Each class has 32 pupils. How many
pupils are there altogether?

3

If I spend
35p a day,
how much do I spend
in a school week?

4

Lunch costs 75p a day.
How much will it
cost me for a week?

5 A batsman played in 29 cricket matches. He scored an average of 47 runs per game.
Calculate the total number of runs he made.

6 A cat eats one tin of Whiskers cat food
every day. A tin costs 29p. What is the
September food-bill for these two cats?

7 The Post Office makes a Christmas Special Offer
of a book of twenty 18p stamps for £3·20.
How much would you save on the usual price of the stamps by buying a book?

8 Tony was out with a can collecting money for a local charity. When his can was emptied
it contained:

Coin	1p	2p	5p	10p	20p	50p	£1
Number	18	27	34	26	16	8	3

How much money did he collect?

9 Which is the cheaper way to buy?
By how much?

HOME COMPUTER

£175 cash, OR
£59 deposit and
52 weekly payments of £3

DIVISION

===== *Exercise 5A   Odd one out* =====

1 Find the odd one out. Give a reason for your answer.

A $8 \div 4$

B $10 \div 5$

C $7 \div 14$

D $18 \div 9$

E $16 \div 8$

2 Which is the odd one out here? Again give a reason for your answer.

A $36 \div 9$

B $28 \div 7$

C $20 \div 5$

D $16 \div 4$

E $12 \div 3$

F $24 \div 4$

G $32 \div 8$

3 Find the odd one out:

    a $24 \div 4$ **b** $54 \div 9$ **c** $30 \div 5$ **d** $18 \div 3$ **e** $35 \div 5$ **f** $48 \div 8$

4 Arrange the following in matching pairs:

A  $56 \div 7$

B $40 \div 8$

C $81 \div 9$

D  $48 \div 6$

E $54 \div 9$

F $45 \div 5$

G $4 \div 16$

H $42 \div 7$

I $35 \div 7$

J $6 \div 24$

5 Arrange these in sets which have the same answer:

    a $64 \div 8$ **b** $36 \div 6$ **c** $54 \div 6$ **d** $72 \div 9$ **e** $49 \div 7$
    f $63 \div 7$ **g** $45 \div 9$ **h** $24 \div 3$ **i** $30 \div 6$ **j** $72 \div 8$

===== *Exercise 5B   Division dominoes* =====

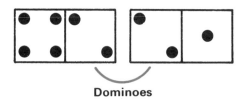

Dominoes

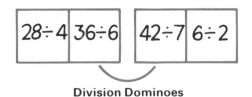

Division Dominoes

1 Fit these together as division dominoes, either as drawings or cut-outs.

 $18 \div 6$ | $7 \div 7$

 $28 \div 7$ | $36 \div 6$

 $25 \div 5$ / $8 \div 8$

 $9 \div 9$ | $16 \div 4$

 $32 \div 8$ | $12 \div 3$

 $36 \div 12$ / $27 \div 9$

 $42 \div 7$ | $15 \div 3$

2 There is something wrong with each of these fittings. Can you find the mistake in each one?

a

10÷1	20÷5

36÷9	40÷8

60÷15	100÷10

b

42÷7	90÷45

84÷42	66÷22

125÷25	42÷14

c

124÷31	96÷12

78÷13	48÷16

33÷11	98÷49

d

55÷11	36÷9

36÷12	16÷4

39÷13	84÷14

3 Arrange the following as they might be after a game of dominoes.

| 216÷72 | | 390÷130 | | 128÷32 | | 165÷33 | 198÷98 | | 54÷18 | | 216÷36 | | 2÷2 | 459÷153 |
| 504÷84 | | 170÷34 | | 243÷81 | | | | 168÷56 | | 288÷72 | | | |

4 Try to make a division domino chain of your own.

MAKING SURE OF DIVISION

(i) **Division by 10, 100, 1000**

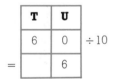

T	U
6	0	÷10

= | | 6 |

1 place to right

H	T	U
3	0	0	÷100

= | | | 3 |

2 places to right

Th	H	T	U
5	0	0	0	÷1000

= | | | | 5 |

3 places to right

Divide by 10, 100, 1000.

Answer is a smaller number.

When a whole number is divided by 10, each figure moves 1 place to the right. Thousands become hundreds, hundreds become tens and tens become units.

=== *Exercise 6* ===

See how quickly and accurately you can do Exercise 6 without using your calculator.

1 Divide by 10:
  a 80 **b** 560 **c** 100 **d** 700 **e** 1040

2 Divide by 100:
  a 600 **b** 900 **c** 12 000 **d** 3800 **e** 14 500

3 Divide by 1000:
  a 3000 **b** 75 000 **c** 80 000 **d** 1000 **e** 1 000 000

4 Calculate:
  a 90÷10 **b** 100÷100 **c** 120÷10 **d** 1000÷10 **e** 2300÷100
  f 58÷2 **g** 93÷3 **h** 72÷4 **i** 435÷5 **j** 366÷6
  k 999÷9 **l** 552÷8 **m** 511÷7 **n** 312÷4 **o** 234 000÷1000
  p 204÷2 **q** 558÷9 **r** 808÷8 **s** 396÷6 **t** 1002÷3

NUMBERS IN ACTION—2

73

OUT AND ABOUT WITH NUMBERS

NUMBERS IN ACTION—2

Exercise 7A

1 There are 84 days until the summer holidays. How many weeks is this?

2 A home economics teacher cuts 65 metres of ribbon into five equal parts. How long is each part?

3 The distance round a school running track is 400 metres. How many laps of the track would there be in a 5000 metres race?

4 A Parents' Association gave the school £200 to buy calculators. If each calculator costs £7, how many could the school buy?

5 182 pupils and 9 teachers are going on a visit to the theatre. They arrange to hire 42-seater buses. How many buses will they need?
Would there be room for any more pupils?

6 John had six shots at winning a prize at the fair. He spent 90p. How much did each shot cost?

7 198 pupils are starting a new school. They will be divided into six classes. How many pupils will be in each class?

8 The leader of a youth club has £1000 to spend on tables and chairs. He buys eight tables at £80 each. He buys chairs at £15 each with the money he has left. How many chairs can he buy?

Exercise 7B

1 How many 18 hole rounds of golf will be played in a 72 hole competition?

2 A dairy sells milk at 26p a pint. A family buys 3 pints every day. Calculate its weekly milk bill.

3 Lampposts are being put up along one side of a straight road 475 metres long. They have to be placed 25 metres apart, with one at each end. How many are needed?

4 In a junior school there are six classes of 33 pupils and four classes of 32 pupils. How many pupils are in the school altogether?

5 Maria is saving up 20p pieces to buy a personal stereo costing £32. She has saved one hundred and thirty-five coins so far. How many more will she need?

6 A school has organised a 15 kilometre sponsored walk. The headmaster is taking part, and 32 people have agreed to sponsor him. Eight sponsors have promised 5p a kilometre, thirteen have promised 10p a kilometre and the rest 20p a kilometre. How much money will the headmaster make for the school if he completes the walk?

7 Pat was helping in the garden, and needed canes to hold up some tall plants. How much would he have to pay for twenty 1000 millimetre canes, fifteen 1500 millimetre canes and twelve 2000 millimetre canes?

8 One Saturday morning Pam and Julie sold flags in the High Street. When their cans were full they went to have their collections counted. How much did each girl collect?

Coins	1p	2p	5p	10p	20p	50p	£1
Pam collected	13	17	23	18	0	6	2
Julie collected	6	23	27	25	3	5	0

=========================== *Exercise 7C* ===========================

1 Is the janitor correct?

2 Bill Smith is a sales representative. His company pays him for the use of his car at these monthly rates:

28p per mile for the first 500 miles.

21p per mile for the next 1000 miles.

14p per mile after that.

One month Bill travelled 1940 miles in his car. How much would he be paid?

How many miles did he travel in a month when he was paid £444·92?

3 Mr Black has 267 sheep to take to the market. A truck can take 56 sheep. How many trucks will he need?

How many more sheep could he take with that number of trucks?

4 A book of stamps used by the local post office contains 200 stamps in each page. The value of the stamps and the number of pages of each type are shown in this table.

Stamp value	1p	2p	5p	10p	15p	18p	22p	30p
Number of pages	3	3	2	4	4	5	5	4

Calculate the total value of the stamps in a new book.
At the end of the day the assistant finds she still has:

Stamp value	1p	2p	5p	10p	15p	18p	22p	30p
Number of stamps left	210	190	230	180	275	264	258	168

How much money did she collect from the sale of stamps?

5 The Simpson family is going by car to Plymouth, a journey of 480 miles. The car's petrol tank holds 50 litres, and the car uses 1 litre every 8 miles.

The tank is full when they leave home.

After 360 miles Mr Simpson stops at a filling station. He asks the attendant for 10 litres of petrol. Was this sensible?

6 A teacher orders ice cream and nougat wafers for the school Christmas Party. There are 176 pupils who will be given one scoop of ice cream and one wafer each. There are 66 wafers in a box and 12 scoops of ice cream in a litre.

How many boxes of wafers and litres of ice cream must the teacher order?

7 Manager: We have 30 000 leaflets to hand out.

Salesman: We could take on forty young people to do the job for us.

Manager: We can pay them 95p for every 50 leaflets they give out—make sure they all get the same number.

How much does each young person earn, and what is the total cost to the company?

PUZZLES

1 A frog is climbing a well 31 feet deep. It climbs 4 feet in one hour, but then slides back 1 foot as it rests for an hour. How long will it take to climb out of the well?

2 Think of a number. Add 3 and multiply your answer by 10. Add 15 and divide your answer by 5. Subtract 9 and divide by 2. What do you notice about the result? Try it again, starting with a different number.

3 Copy this diagram in your notebook. Put whole numbers in the boxes so that each answer is 54. A number must not be used more than once. Try it again, without using any of the previous numbers.

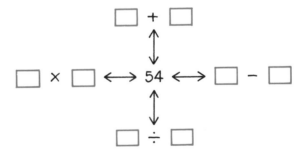

4 Starting with 501 at darts, what is the smallest number of darts you must throw to reach zero?

5 Think of a number. Double it and add 4. Multiply by 6 and subtract 18. Divide by 3, subtract 2 and divide by 4. How does your answer compare with the number you first thought of?

6 If you fill in these boxes correctly, each calculation will contain the digits 1, 2, 3, 4, 5, 6, 7, 8, 9.

a ☐ 2 × 48 ☐ = 5 ☐ 96   **b** 3 ☐ × 18 ☐ = 72 ☐ 4

7 Surprise your friends by saying that you can add all the whole numbers from 1 to 1000 in less than ten seconds! How can you do this? Start by adding the first and last numbers. One multiplication will give you the answer. Think the rest out.
Test your method for $1 + 2 + 3 + \ldots + 8 + 9 + 10$.

NUMBERS IN ACTION—2

A palindrome number reads the same backwards and forwards, for example 1991.

Choose a number greater than 9.   39
Reverse its digits.   93
                                   ⸻
Add   132
If not a palindrome, reverse it   231
                                   ⸻
Add   363, a palindrome.

Tens

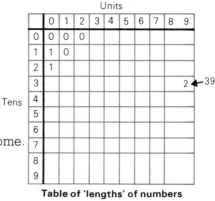

	Units									
	0	1	2	3	4	5	6	7	8	9
0	0	0	0							
1	1	0								
2	1									
3										2 ◄—39
4										
5										
6										
7										
8										
9										

Table of 'lengths' of numbers

There were two additions, so 39 has 'length' 2.
Try this with other numbers. Some will take a long time to turn into palindromes.
Copy and complete the table of 'lengths' of numbers.
You may need some of your neighbours to help you.
You will find some interesting patterns of numbers in the table.

CHECK-UP ON **NUMBERS IN ACTION—2**

1A   **Multiplying and dividing whole numbers**

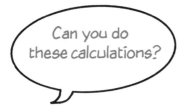

Can you do
these calculations?

a 63×8

b 37×26

c $342 \div 9$

d 13×0

e 1×987

f Multiply 84 by 10, by 100, and by 1000.

g Divide 3700 by 10, by 100 and by 1000.

h Divide a million by a thousand.

2A, B Problems with whole numbers

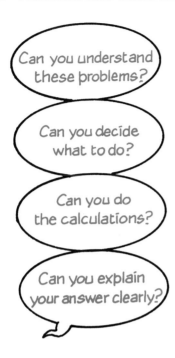

a A gardener pays £56 for 14 slabs for his path.
How much does each slab cost?
How much would he pay for 18 slabs?

b Pens can be bought for 18p each, or 90p for 6.
Describe the cheapest and dearest ways of buying 20 pens.
What is the difference in cost between these two ways?

c A carton of crisps contains 40 bags. 290 bags were sold. How many cartons had to be opened?

d A sheet of 31p stamps consists of 15 rows with 12 stamps in each row. What is its total value?

3A, B, C Number sequences

a 9, 27, 81, ... a fourth number.

b 1000, 200, 40, ... a fourth number.

c 1, 2, 4, 8, ... The first number to be greater than 1000.

INTRODUCING LETTERS

Empty the bags

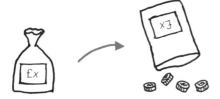

This bag has
x £1 coins in it.

Empty the bag,
and count the coins.

There were 4 coins in the bag,
so $x = 4$.

=== *Exercise 1* ===

Find the number each letter stands for. Give your answer like this: $x = 4$.

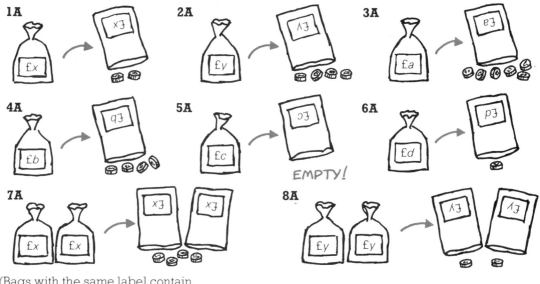

1A   **2A**   **3A**

4A   **5A**   **6A**

EMPTY!

7A   **8A**

(Bags with the same label contain
the same number of coins.)

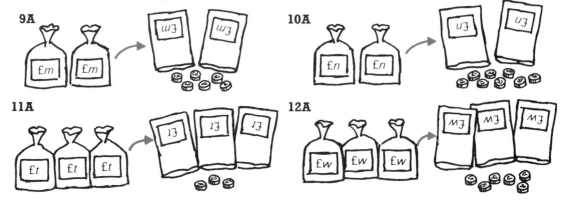

9A   **10A**

11A   **12A**

13A

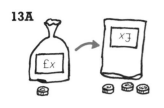

(Careful! There was one there already.)

14A

15A

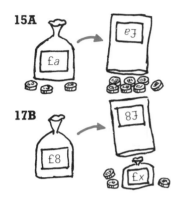

17B

(There is another bag in this bag!)

16A

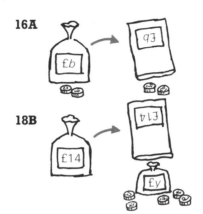

18B

19B

20B

21B

23C

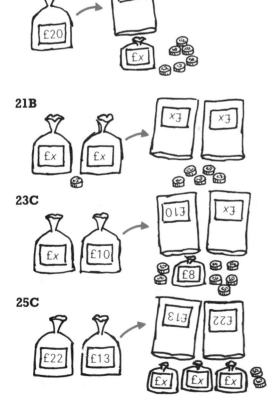

22B

24C

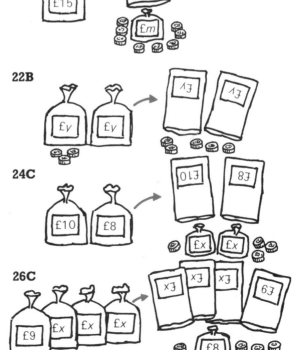

25C

26C

Open bags and burst bags

This bag contains 12 apples.

It bursts and 2 apples fall out. 10 are left.

The bag is opened and 3 more apples are dropped in. There are now 13 apples in the bag.

x apples

$x-2$ apples

$x+1$ apples

=== *Exercise 2* ===

Write down the number of apples in each right-hand bag.

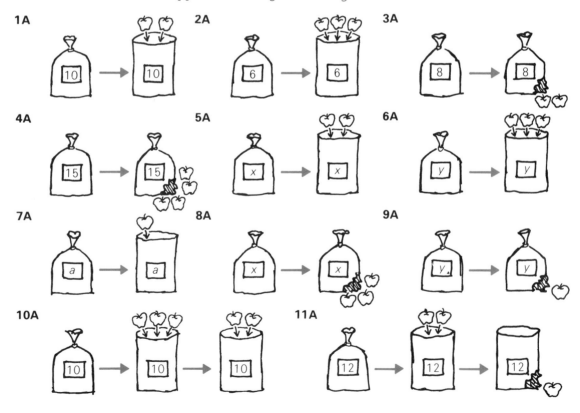

1A

2A

3A

4A

5A

6A

7A

8A

9A

10A

11A

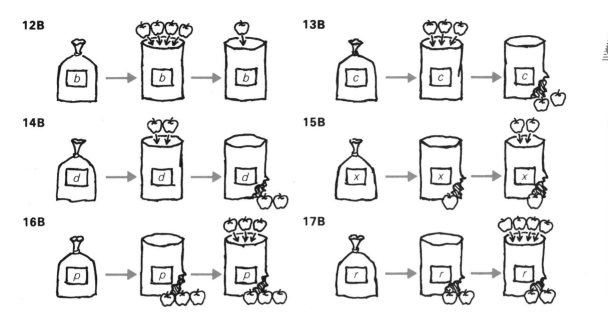

12B **13B** **14B** **15B** **16B** **17B**

Apple carts

How many apples are in these carts?

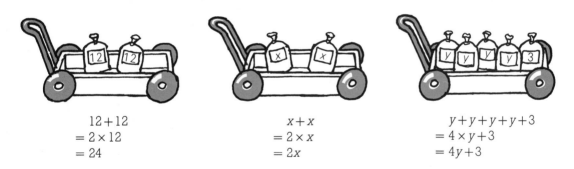

$12 + 12$
$= 2 \times 12$
$= 24$

$x + x$
$= 2 \times x$
$= 2x$

$y + y + y + y + 3$
$= 4 \times y + 3$
$= 4y + 3$

===== *Exercise 3* =====

Write down expressions for the number of apples in these carts.

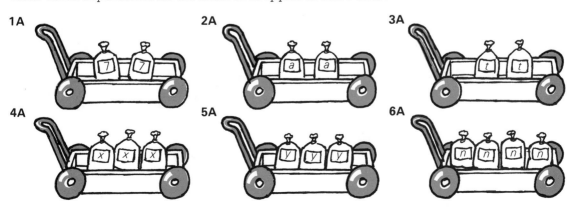

1A **2A** **3A** **4A** **5A** **6A**

LETTERS FOR NUMBERS

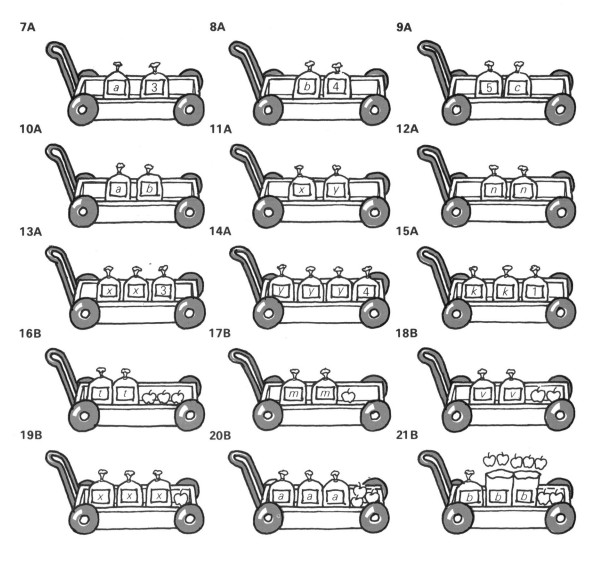

Use all the information in the pictures to find expressions for the number of apples in these carts:

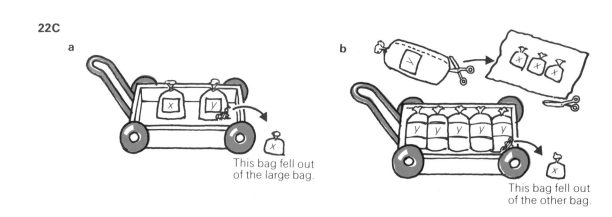

22C

a

This bag fell out of the large bag.

b

This bag fell out of the other bag.

84

Target Practice

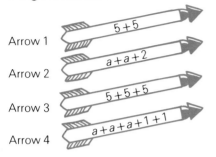

Arrow 1: $5+5$
Arrow 2: $a+a+2$
Arrow 3: $5+5+5$
Arrow 4: $a+a+a+1+1$

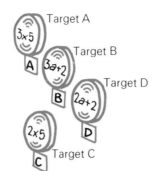

Target A: 3×5
Target B: $3a+2$
Target D: $2a+2$
Target C: 2×5

Matching arrows
with targets:

Arrows		Targets
1	$\longrightarrow$	C
2	$\longrightarrow$	D
3	$\longrightarrow$	A
4	$\longrightarrow$	B

Exercise 4

Match the arrows with the targets. Show the matching in a table, as above:

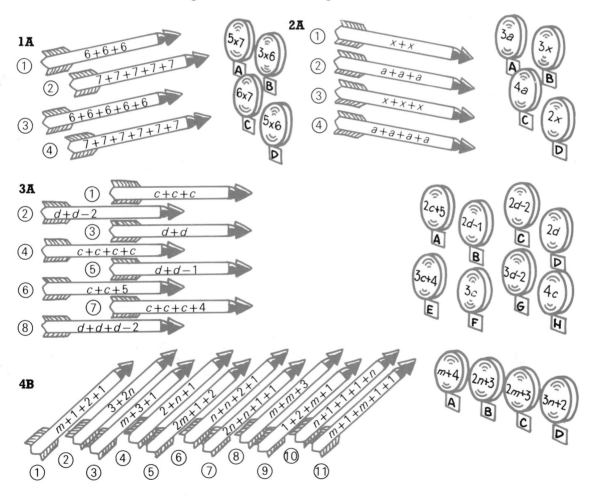

1A
① $6+6+6$
② $7+7+7+7+7$
③ $6+6+6+6+6$
④ $7+7+7+7+7+7$

A: 5×7
B: 3×6
C: 6×7
D: 5×6

2A
① $x+x$
② $a+a+a$
③ $x+x+x$
④ $a+a+a+a$

A: $3a$
B: $3x$
C: $4a$
D: $2x$

3A
① $c+c+c$
② $d+d-2$
③ $d+d$
④ $c+c+c+c$
⑤ $d+d-1$
⑥ $c+c+5$
⑦ $c+c+c+4$
⑧ $d+d+d-2$

A: $2c+5$
B: $2d-1$
C: $2d-2$
D: $2d$
E: $3c+4$
F: $3c$
G: $3d-2$
H: $4c$

4B
① $m+1+2+1$
② $3+2n$
③ $m+3+1$
④ $2+n+1$
⑤ $2m+1+2$
⑥ $n+n+2+1$
⑦ $m+m+3$
⑧ $2n+n+1+1$
⑨ $1+2+m+1$
⑩ $n+1+1+1+n$
⑪ $m+1+m+1+1$

A: $m+4$
B: $2n+3$
C: $2m+3$
D: $3n+2$

Careful! Some arrows missed their targets, and several arrows hit the same target.

Making sure

Examples: 1 $4+4+4+4+4 = 5 \times 4 = 20$
2 $a+a+a = 3 \times a = 3a$
3 $x+x+4-1 = 2x+3$

=========== *Exercise 5* ===========

Write these in a shorter form.

1A $7+7$ **2A** $6+6+6$ **3A** $1+1+1+1+1$ **4A** $9+9+9+9$
5A $x+x$ **6A** $y+y$ **7A** $a+a+a$ **8A** $b+b+b+b$
9A $c+c+c$ **10A** $d+d$ **11A** $t+t+t+t$ **12A** $k+k+k+k+k$
13A $x+x+5$ **14A** $y+y-1$ **15A** $m+m+m+4$ **16A** $n+n-2$
17A $t+t+t+3$ **18A** $v+v-4$ **19A** $a+a+a+a+1$ **20A** $b+b+b-3$
21A $x+2+3$ **22A** $y+4+1$ **23A** $k+3-1$ **24A** $n+2+2$
25A $2x+3+4$ **26A** $3y+2-2$ **27A** $4t-1+2$ **28A** $5a-2+4$
29B $x+x+3+1$ **30B** $y+y+2-1$ **31B** $p+2+p+1$ **32B** $r+1+r-1+r$
33B $3+x-2+x$ **34B** $4+a-4+a$ **35B** $2a+2a+2a$ **36B** $5b+5b+5b+5b$

PATTERNS

Starting patterns

1 fork
3 prongs

2 forks
6 prongs

3 forks
9 prongs

• • •

In a table,

Number of forks	1	2	3	4
Number of prongs	3	6	9	12

The number of prongs is always 3 times the number of forks. So n forks would have $3n$ prongs.
Remember: $3n$ means $3 \times n$.

=========== *Exercise 6* ===========

1A

1 plant
2 leaves

 2 plants

   3 plants

Copy and complete this table.

Number of plants	1	2	3	4	...	100	...	n
Number of leaves								

How many leaves do 100 plants have?
How many leaves do n plants have?

2A

Make up a table showing the number of legs 1, 2, 3, 4, ..., 100, ..., *n* dogs have. How many legs for 50 dogs? For *x* dogs?

3A

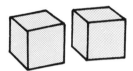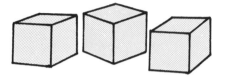

Make up a table showing the number of faces that 1, 2, 3, 4, ..., 100, ..., *n* cubes have. How many faces do 50 cubes have? How many do *k* cubes have?

4A Do question **3A** again, but this time for the number of corners.

5A Do question **3A** again, but this time for the number of edges.

6A

Make a table showing the number of fingers on 1, 2, 3, 4, ..., *n* hands.

7B

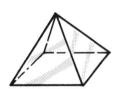

Triangular pyramid   Square pyramid   Pentagonal pyramid

Construct a table showing the total number of edges of pyramids with 3, 4, 5, ... edges for their bases.
How many edges has a pyramid with an *n*-sided base?

8B Repeat question **7B** to investigate the number of faces of a pyramid with an *n*-sided base.

LETTERS FOR NUMBERS

9B

2 coins

1 pair touching

3 coins

2 different
pairs touching

4 coins

Make a table to show the number of different pairs of coins touching for 2 coins, 3 coins, 4 coins, ..., 100, ..., n coins.

How many such pairs are there for y coins?

10C a Investigate the number of different pairs of touching five-pence coins in these patterns:

b Investigate other patterns of coins.

Newspaper patterns

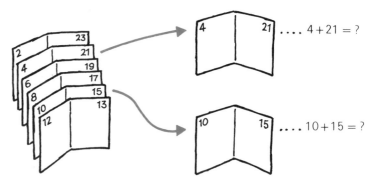

.... $4 + 21 = ?$

.... $10 + 15 = ?$

Look at a newspaper to see if its pages are numbered like those above.

Exercise 7

1A What do you notice when you add the numbers on facing pages of the same sheet as shown above?

2A Write down the numbers which these letters stand for.

3A What is the page number on the back of page 2?

4A Find the value of x in each of these:

(the back of page 12)

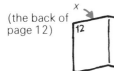  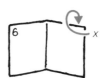

5B Find numbers to replace the letters in these:

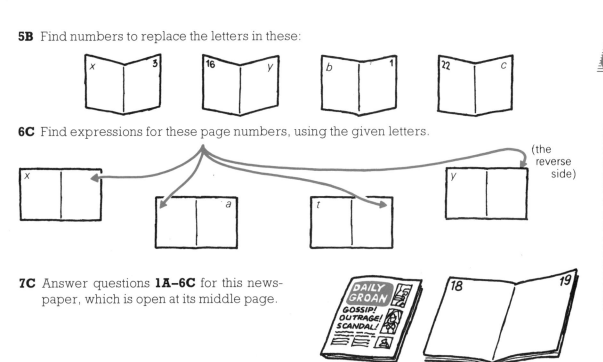

6C Find expressions for these page numbers, using the given letters.

(the reverse side)

7C Answer questions **1A–6C** for this newspaper, which is open at its middle page.

DAILY GROAN
GOSSIP!
OUTRAGE!
SCANDAL!

18   19

Number patterns

=== *Exercise 8* ===

Copy and complete these tables:

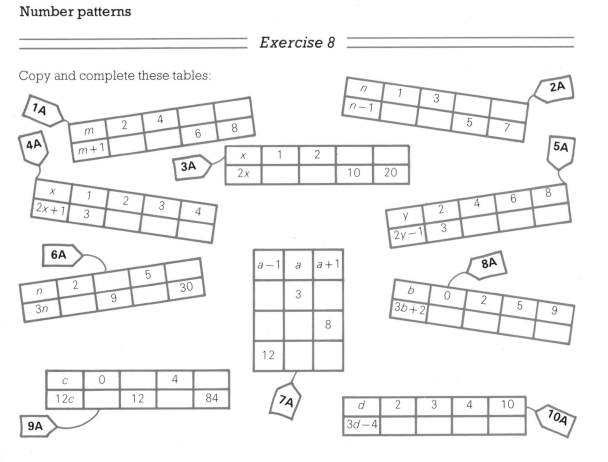

1A

m	2	4		6	8
m+1					

2A

n	1	3			
n−1				5	7

3A

x	1	2		10	20
2x					

4A

x	1	2	3		4
2x+1	3				

5A

y		2	4	6	8
2y−1	3				

6A

n	2		5		
3n		9		30	

7A

a−1	a	a+1
	3	
		8
12		

8A

b	0	2	5	9
3b+2				

9A

c	0		4	
12c		12		84

10A

d	2	3	4	10
3d−4				

89

LETTERS FOR NUMBERS

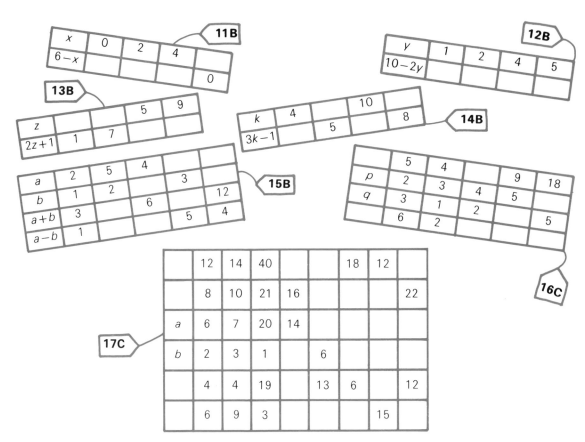

11B

x	0	2	4	
6 − x				0

12B

y		1	2	4	5
10 − 2y					

13B

z			5	9
2z + 1	1	7		

14B

k	4			10	
3k − 1		5		8	

15B

a	2	5	4		3	
b	1	2		6		12
a + b	3				5	4
a − b	1					

16C

p	2	3			9	18
q	3	1	4	5		
	6	2				5

17C

	12	14	40			18	12	
	8	10	21	16				22
a	6	7	20	14				
b	2	3	1		6			
	4	4	19		13	6		12
	6	9	3				15	

Perimeter patterns

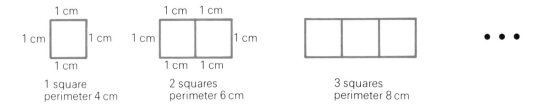

1 cm

1 cm ⬜ 1 cm

1 cm

1 square
perimeter 4 cm

1 cm 1 cm

1 cm ⬜⬜ 1 cm

1 cm 1 cm

2 squares
perimeter 6 cm

3 squares
perimeter 8 cm

• • •

How can we calculate the perimeter of the rectangle from the number of squares?

Number of squares	1	2	3	4	…	n
Perimeter in cm	4	6	8	10		?

② × number of squares	2	4	6	8	…	2n
② × number of squares + 2	4	6	8	10		2n + 2

For n squares the perimeter in centimetres is $2n + 2$.

In this exercise the lengths are in centimetres.

1A 　•　•　•

Copy and complete this table for the patterns of triangles above.

Number of triangles	1	2	3	4	5	6	...	10	...	n
Perimeter in cm										

2A   　•　•　•

Copy and complete this table for the patterns of triangles above.

Number of triangles	1	2	3	4	5	6	...	10	...	n
Perimeter in cm										

3A   　•　•　•

Make a table for the number of rectangles and the number of cm in their perimeters.
By how much do the perimeters increase each time?
What is the perimeter of the figure made from n rectangles?
Check your answer by trying it for 1, 2, 3, rectangles.

4A  　•　•　•

Make a table for the number of 'house' shapes above and the number of cm in their perimeters.
By how much do the perimeters increase each time?
What is the perimeter of the figure formed by n shapes?

5B 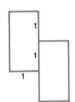 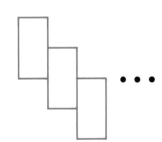　•　•　•

Investigate the rule for finding the perimeters of these shapes.

6C Repeat question **5B** for 3 by 1 rectangles which overlap by: **a** 2 cm **b** 1 cm.

7C Repeat question **5B** for 4 by 1 rectangles which overlap by a whole number of centimetres.
Can you find four different expressions for the perimeters of the figures with n rectangles?

EQUATIONS

A cover up

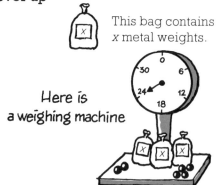

This bag contains *x* metal weights.

Here is a weighing machine

Each metal weight weighs 1 kg.

The 3 bags of weights and the 6 weights outside weigh 24 kg.

So $x+x+x+6 = 24$

or $3x+6 = 24$

This is called an equation.

To solve the equation $3x+6 = 24$, find the number that *x* stands for.

$3x+6 = 24$ 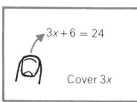 Cover 3x	 $+6 = 24$ What number and six equals twenty-four?	 Eighteen and six gives twenty-four
$3x+6 = 24$ So 3x is the same as eighteen	$3x = 18$  Cover x	$3\ \square = 18$ 3 whats are eighteen?
$3\ \boxed{6} = 18$ 3 sixes are eighteen	$3x = 18$ So x is the same as six	$x = 6$

$x = 6$ is the solution of the equation.
There are 6 weights in each bag on the weighing machine.

=========================== *Exercise 10* ===========================

Use this 'cover up' method to solve these equations.

1A $x+4 = 7$ **2A** $y-2 = 1$ **3A** $z+5 = 10$

4A $2x+1 = 5$ **5A** $2x+7 = 9$ **6A** $2x+3 = 15$

7A $2y+3 = 3$ **8A** $2x+3 = 27$ **9A** $2n-1 = 5$

10A $3m-4 = 14$ **11A** $3p-7 = 2$ **12A** $3q+6 = 15$

13A $4x+1 = 25$ **14A** $4y-2 = 10$ **15A** $4z-4 = 0$

16A $1+2a = 9$ **17A** $3+6b = 9$ **18A** $5+8c = 5$

19A $12+4d = 28$ **20A** $10x-20 = 80$ **21A** $11y-23 = 32$

Solve the equation $48-7p = 13$.

$$48 - 7p = 13 \qquad 48 - \boxed{35} = 13$$

$$7p = 35 \qquad 48 - 7p = 13$$
$$\boxed{35}$$

$$p = 5$$

22B $15-4x = 3$ **23B** $20-3y = 5$ **24B** $47-6p = 5$

25B $8-2x = 4$ **26B** $9 = 2a-1$ **27B** $31 = 2b+15$

28B $27 = 3+4w$ **29B** $9 = 5+4y$ **30B** $0 = 36-4k$

31B $19 = 11+2p$ **32B** $17 = 3q-4$ **33B** $6r-13 = 17$

Weighing bags

The 2 bags of weights and the 3 weights outside weigh 9 kg.
This makes the equation:

$$2y + 3 = 9$$

$$\boxed{6} + 3 = 9$$

$$2y = 6$$

$$y = 3$$

$y = 3$ is the solution of the equation.
There are 3 weights in each bag.

═══════════════ *Exercise 11* ═══════════════

Make an equation for each of the following, and solve it. Then say how many weights there are in each bag.

1A

2A

3A

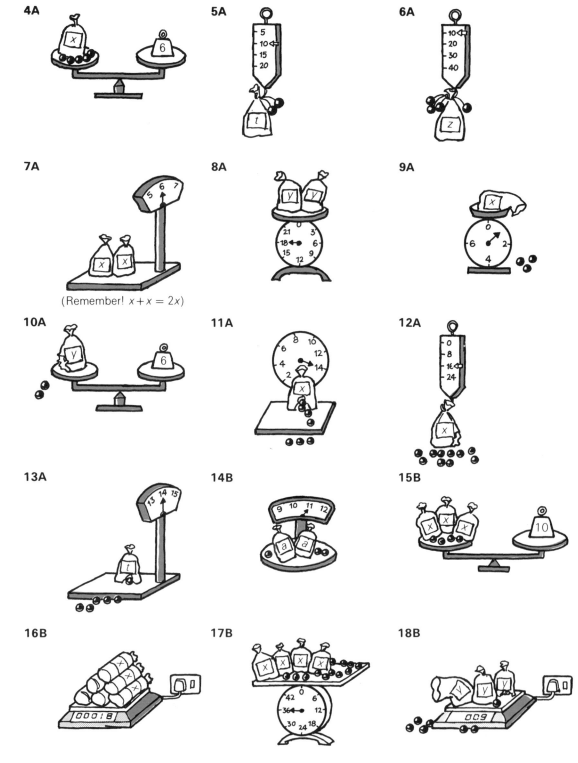

4A

5A

6A

7A

(Remember! $x + x = 2x$)

8A

9A

10A

11A

12A

13A

14B

15B

16B

17B

18B

19B

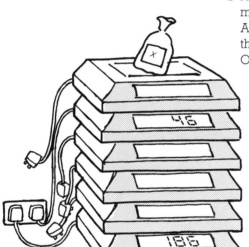

20C

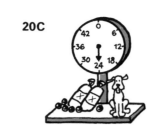

Use information from both pictures to find what number x stands for. What is the weight of the dog?

1 Here is a tower of identical weighing machines.

A weighing machine will weigh everything that is piled on top of it.

Only two of these machines are working.

How many 1 kg weights are in the bag labelled x?

2 These two pictures show the same weighing machine on top of another weighing machine. How many 1 kg weights are in the bag labelled x?

Warning! Could cause brain strain!

Which key?

Here is a calculator puzzle.

Keys pressed: $\boxed{?}\ \boxed{+}\ \boxed{4}\ \boxed{=}$ What is the key that was pressed first?

Display   : $\boxed{\qquad\qquad 5}$

Using x for the number on the unknown key,

$$x + 4 = 5$$

$x\ = 1$ is the solution of this equation.

So the $\boxed{1}$ key was pressed first.

=========================== *Exercise 12* ===========================

In each question use x for the number on the unknown key. Then make an equation, and solve it.
Check the solution with your calculator.

1A

$\boxed{?}\ \boxed{+}\ \boxed{4}\ \boxed{=}$
$\boxed{\qquad\qquad 9}$

2A

$\boxed{?}\ \boxed{-}\ \boxed{3}\ \boxed{=}$
$\boxed{\qquad\qquad 3}$

3A

$\boxed{2}\ \boxed{+}\ \boxed{?}\ \boxed{=}$
$\boxed{\qquad\qquad 10}$

4A

$\boxed{9}\ \boxed{-}\ \boxed{?}\ \boxed{=}$
$\boxed{\qquad\qquad 8}$

5A

$\boxed{?}\ \boxed{\times}\ \boxed{9}\ \boxed{=}$
$\boxed{\qquad\qquad 72}$

6A

$\boxed{6}\ \boxed{8}\ \boxed{\div}\ \boxed{?}\ \boxed{=}$
$\boxed{\qquad\qquad 17}$

7A

$\boxed{?}\ \boxed{+}\ \boxed{?}\ \boxed{=}$
$\boxed{\qquad\qquad 10}$

In **7A** the equation is
$$x + x = 10$$

8A

$\boxed{?}\ \boxed{+}\ \boxed{?}\ \boxed{=}$
$\boxed{\qquad\qquad 16}$

9A

$\boxed{?}\ \boxed{\times}\ \boxed{?}\ \boxed{=}$
$\boxed{\qquad\qquad 81}$

10A $\boxed{2}\ \boxed{\times}\ \boxed{?}\ \boxed{+}\ \boxed{1}\ \boxed{=}$

$\boxed{\qquad\qquad 5}$

11A $\boxed{1}\ \boxed{2}\ \boxed{8}\ \boxed{-}\ \boxed{?}\ \boxed{=}$

$\boxed{\qquad\qquad 119}$

12A $\boxed{1}\ \boxed{0}\ \boxed{-}\ \boxed{2}\ \boxed{\times}\ \boxed{?}\ \boxed{=}$

$\boxed{\qquad\qquad 8}$

13A $\boxed{3}\ \boxed{\times}\ \boxed{?}\ \boxed{+}\ \boxed{2}\ \boxed{=}$

$\boxed{\qquad\qquad 8}$

14B $\boxed{?}\ \boxed{\times}\ \boxed{?}\ \boxed{+}\ \boxed{1}\ \boxed{=}$

$\boxed{\qquad\qquad 5}$

15B $\boxed{?}\ \boxed{\times}\ \boxed{2}\ \boxed{+}\ \boxed{3}\ \boxed{=}$

$\boxed{\qquad\qquad 9}$

16B $\boxed{?}\ \boxed{\times}\ \boxed{5}\ \boxed{-}\ \boxed{5}\ \boxed{=}$

$\boxed{\qquad\qquad 0}$

17B $\boxed{?}\ \boxed{\times}\ \boxed{3}\ \boxed{-}\ \boxed{4}\ \boxed{=}$

$\boxed{\qquad\qquad 20}$

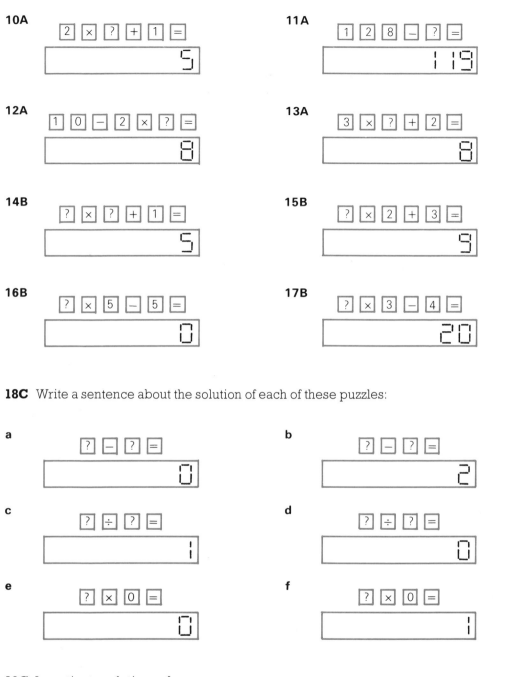

18C Write a sentence about the solution of each of these puzzles:

a $\boxed{?}\ \boxed{-}\ \boxed{?}\ \boxed{=}$

$\boxed{\qquad\qquad 0}$

b $\boxed{?}\ \boxed{-}\ \boxed{?}\ \boxed{=}$

$\boxed{\qquad\qquad 2}$

c $\boxed{?}\ \boxed{\div}\ \boxed{?}\ \boxed{=}$

$\boxed{\qquad\qquad 1}$

d $\boxed{?}\ \boxed{\div}\ \boxed{?}\ \boxed{=}$

$\boxed{\qquad\qquad 0}$

e $\boxed{?}\ \boxed{\times}\ \boxed{0}\ \boxed{=}$

$\boxed{\qquad\qquad 0}$

f $\boxed{?}\ \boxed{\times}\ \boxed{0}\ \boxed{=}$

$\boxed{\qquad\qquad 1}$

19C Investigate solutions of:

$\boxed{1}\ \boxed{2}\ \boxed{\div}\ \boxed{?}\ \boxed{=}$

$\boxed{\qquad\qquad}$

$\boxed{3}\ \boxed{6}\ \boxed{0}\ \boxed{\div}\ \boxed{?}\ \boxed{=}$

$\boxed{\qquad\qquad}$

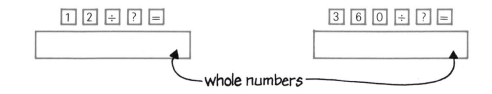

whole numbers

A return visit

On page 86 you discovered that n plants have $2n$ leaves. If there were 64 leaves, then $2n$ would be the same as 64.

$$\text{So} \quad 2n = 64$$
$$n = 32$$

There were 32 plants.

=========================== *Exercise 13* ===========================

In each question make an equation and then solve it.

1A n forks have $3n$ prongs. How many forks have 45 prongs altogether?

2A m dogs have $4m$ legs. How many dogs have 136 legs altogether?

3A k cubes have **a** $6k$ faces. How many cubes are there for 72 faces?
b $8k$ corners. ,, ,, ,, ,, ,, ,, 40 corners?
c $12k$ edges. ,, ,, ,, ,, ,, ,, 96 edges?

4A **a** A pyramid with p sides in its base has $2p$ edges. How many sides are in the base of a pyramid which has 72 edges?
b The pyramid has $p+1$ faces. How many sides are in the base of a pyramid which has 15 faces?

5A The perimeter of an arrangement of q squares like this is $2q+2$ cm. How many squares are in this arrangement with perimeter 48 cm?

1 cm

1 cm   1 cm   1 cm

6A 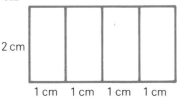 The perimeter of an arrangement of r rectangles like this is $2r+4$ cm.
How many rectangles are in the arrangement with perimeter 30 cm?

2 cm

1 cm   1 cm   1 cm   1 cm

7B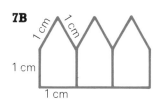

Show that the perimeter of a figure with x shapes like this is $3x+2$ cm.
How many shapes are in the figure with perimeter 38 cm?

8B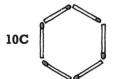

Show that the perimeter of a figure with y rectangles arranged as shown is $4y+2$ cm.
How many rectangles are in the figure with perimeter 58 cm?

9C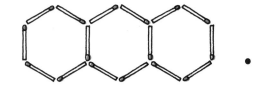

These shapes are made of matchsticks.
a How many triangles are there in the shape with a *perimeter* made up of 200 matches?
b How many triangles are there in the *shape* which consists of 151 matches?

10C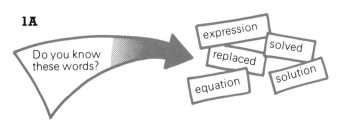

These shapes are made of matchsticks.
a How many hexagons are there in the shape with a *perimeter* made up of 346 matches?
b How many hexagons are there in the *shape* which consists of 1001 matches?

CHECK-UP ON **LETTERS FOR NUMBERS**

1A

Do you know these words?

expression
solved
replaced
equation
solution

Copy and complete:

(i) $4x+2$ is an _____. The letter x can be _____ by a number.

(ii) $4x+2=10$ is an _____. It can be _____ to give the _____ $x=2$.

LETTERS FOR NUMBERS

2A, B

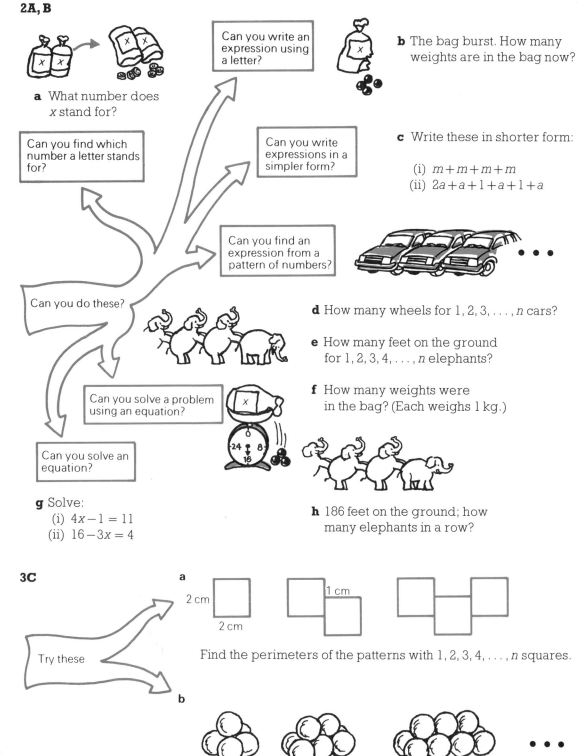

a What number does x stand for?

Can you write an expression using a letter?

b The bag burst. How many weights are in the bag now?

Can you find which number a letter stands for?

Can you write expressions in a simpler form?

c Write these in shorter form:

(i) $m+m+m+m$
(ii) $2a+a+1+a+1+a$

Can you find an expression from a pattern of numbers?

Can you do these?

d How many wheels for $1, 2, 3, \ldots, n$ cars?

e How many feet on the ground for $1, 2, 3, 4, \ldots, n$ elephants?

f How many weights were in the bag? (Each weighs 1 kg.)

Can you solve a problem using an equation?

Can you solve an equation?

g Solve:
(i) $4x-1=11$
(ii) $16-3x=4$

h 186 feet on the ground; how many elephants in a row?

3C

a

2 cm 1 cm 2 cm

Try these

Find the perimeters of the patterns with $1, 2, 3, 4, \ldots, n$ squares.

b

Find the number of snowballs in the nth pile.

RECOGNISING AND NAMING ANGLES

Class discussion—Talk about angles

back angle

front angle

The boy's arms make angles. They make a smaller 'front' angle and a larger 'back' angle.

1A Where can you see angles in the pictures below?

2A Now can you see angles in the classroom?

 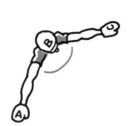 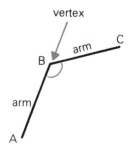

The smaller 'front' angle is named ABC. You can also call it CBA. It is written ∠ABC or ∠CBA. ∠ means 'angle'. We will not use the larger 'back' angle in this chapter.

========= *Exercise 1* =========

1A Name the angles in these pictures.

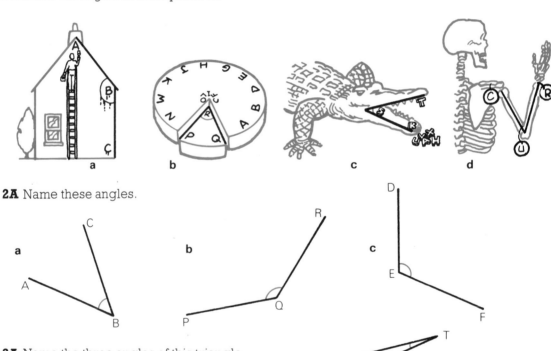

a   b   c   d

2A Name these angles.

a   b   c

3A Name the three angles of this triangle.

4A a Name three angles which have NM as an arm.
  b Name all the angles which have KM as an arm.

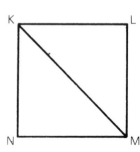

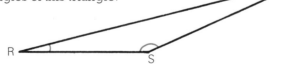

An open book

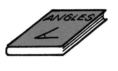

The book is opened

A page is raised

=== *Exercise 2* ===

1A Name two angles that have DB as an arm.

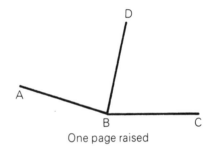

One page raised

2A Name all the angles that have the new page
EB as an arm.
There are more than two.
How many are there?

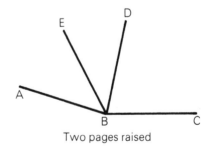

Two pages raised

3A Name all the angles that have the new page
FB as an arm.
How many are there?

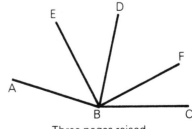

Three pages raised

4A Copy and complete this table which shows the number of pages raised and the number of angles that have the new page as an arm.

Number of pages raised	1	2	3
Number of 'new page' angles			

5B Copy the diagram of question **3A**. Add to it a fourth page between AB and EB. Call it GB. Name all the angles that have this new page as an arm.

6B Describe how to calculate the number of 'new page' angles if you know the number of pages raised. You may need to add to your table for question **4A** to help you find the rule.

7C Shut the Angles Book. This time when you open it and start to raise pages name *all* the angles in the diagram, not just the ones with the new page as an arm.

    a Make up a table of your results.

    b Can you suggest the number of angles formed when 10 pages are raised? 50 pages?

    c With *n* pages raised how many angles are there?

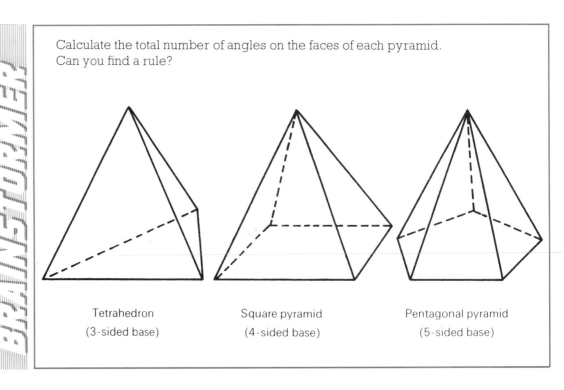

Calculate the total number of angles on the faces of each pyramid.
Can you find a rule?

| Tetrahedron | Square pyramid | Pentagonal pyramid |
| (3-sided base) | (4-sided base) | (5-sided base) |

FITTING ANGLES AROUND A POINT

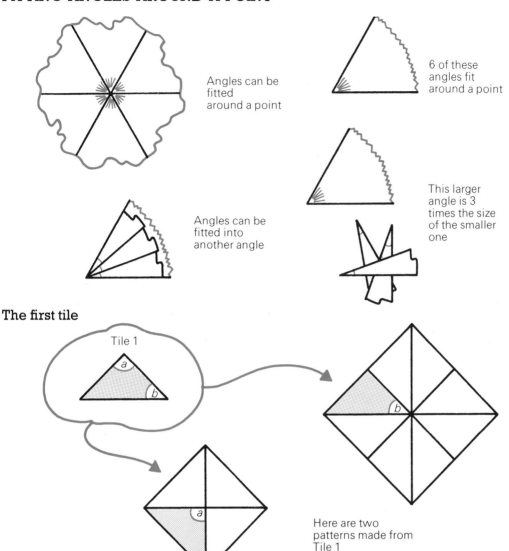

Angles can be fitted around a point

6 of these angles fit around a point

Angles can be fitted into another angle

This larger angle is 3 times the size of the smaller one

The first tile

Tile 1

Here are two patterns made from Tile 1

Exercise 3

1A On squared paper draw two squares with sides 4 cm long.
Draw both diagonals (from corner to corner) in each one.
Cut out the squares, and then cut along the diagonals.
You should now have eight tiles like Tile 1 above.

2A a How many of angle *a* fit around a point?
   b How many of angle *b* fit around a point?

3A Copy and complete this table:

Angle	a	b
Number fitting around a point	*	*

4A Using your cut-out tiles, or the patterns above, or the table, find how many of angle b would fit into angle a.

5A Copy and complete: Angle a is ___* times the size of angle b.

6B How are the three starred numbers linked together?

The second tile

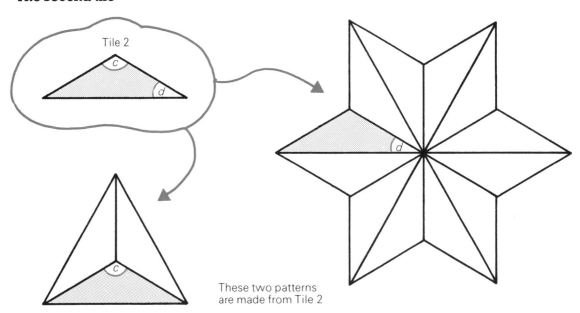

Tile 2

These two patterns are made from Tile 2

═══════════════ *Exercise 4* ═══════════════

1A **a** How many of angle c fit around a point?
  b How many of angle d fit around a point?

2A Copy and complete this table.

Angle	c	d
Number fitting around a point	*	*

3A How many of angle d would fit into angle c?

4A Copy and complete: Angle c is ___* times the size of angle d.

5B Are the three starred numbers connected in the same way as for Tile 1?

The third tile

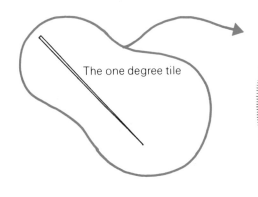

Tile 3

Here is a pattern made from Tile 3.

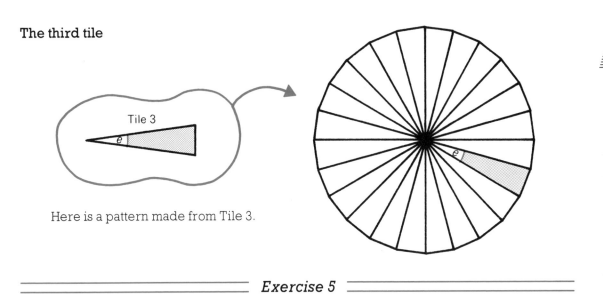

===== *Exercise 5* =====

1A How many of angle e fit around a point?

2A Copy and complete this table:

Angle	e	a	b	c	d
Number fitting around a point	*	*	*	*	*

3B How many of angle e would fit into each of the angles a, b, c and d?

4B Explain clearly the connections between the starred numbers.

DEGREES AND PROTRACTORS

Protractor patterns

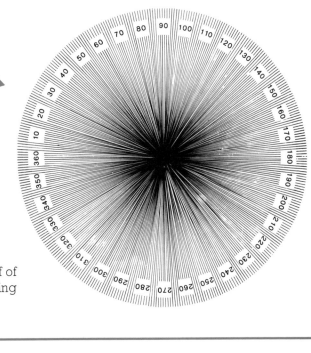

The one degree tile

Look at your own protractor. It shows half of this pattern, but has some of the lines missing to make it easier to use.

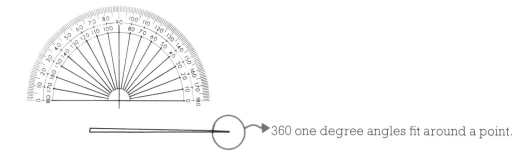

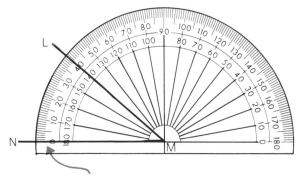

360 one degree angles fit around a point.

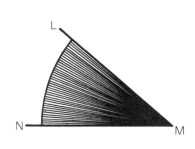

This **outside scale** can be used to count the number of one degree angles that fit into ∠LMN.

40 one degree angles fit into ∠LMN.

∠LMN is a 40 degree angle.
∠LMN is a 40° angle.
∠LMN = 40°.

Exercise 6

1A Write down the size in degrees of each angle, like this: ∠LMN = 40°.

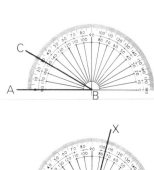

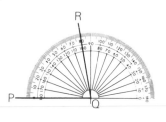

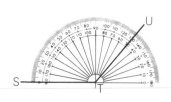

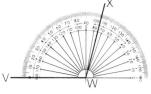

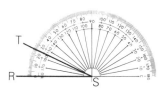

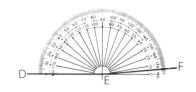

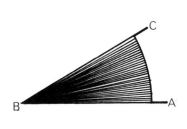

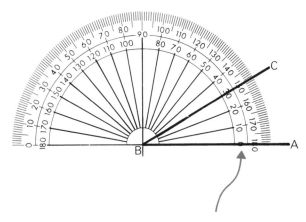

30 one degree angles fit into ∠ ABC.

This **inside scale** can be used to count the number of one degree angles that fit into ∠ ABC.

∠ ABC is a 30 degree angle.
∠ ABC is a 30° angle.
∠ ABC = 30°.

2A Write down the size in degrees of each of these angles.

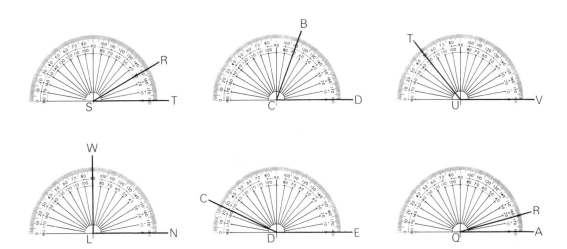

3A Guess the size of each angle. Then measure it with a protractor. Write your answers like this $\angle XYZ = 43°$.

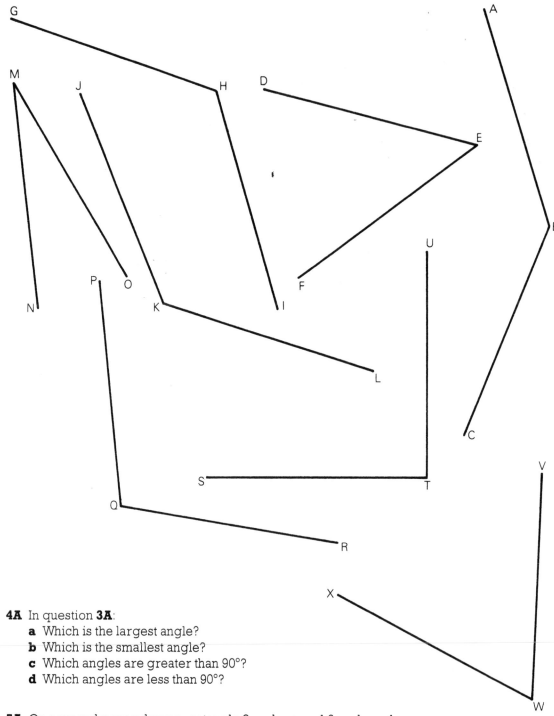

4A In question **3A**:
  a Which is the largest angle?
  b Which is the smallest angle?
  c Which angles are greater than 90°?
  d Which angles are less than 90°?

5A On squared paper draw a rectangle 8 cm long and 6 cm broad.
Draw one diagonal (from corner to corner).
Measure all the angles you can find in your drawing.

6A Measure ∠ ABD and ∠ DBC in the picture of the 50p coin. Which is the larger of the two angles?

7A Find the sum of the two angles in question **6A**. If ∠ ABC is thought of as a **straight angle**, what size is it in degrees?

8A Name and measure the three angles of triangle PQR. Find the sum of these three angles.

9A Draw a large triangle, and mark its angles as shown. Cut out the triangle, and tear along the dotted lines. Place the three marked angles together. What do you find?

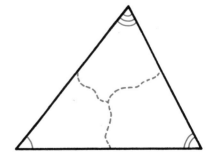

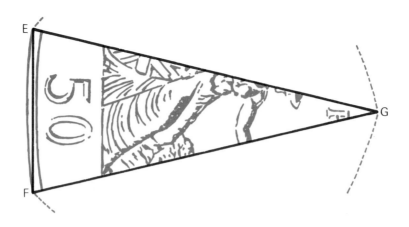

10A Measure all the angles of triangle EFG. Find their sum.

11B **a** What do your answers to questions **8A**, **9A** and **10A** suggest about the sum of the three angles in a '50p triangle'?

b Using a tracing of the 50p coin diagram on page 111 find a different '50p triangle'.

c Use this triangle to check what you found in **a**.

d Is this rule only true for '50p triangles'?

Investigate by drawing some other triangles and measuring their angles using a protractor.

12C Investigate the sum of the angles of a quadrilateral (four-sided figure).

What about pentagons, hexagons, heptagons, etc?

Why 360?

Here is Laura Stuart's report on her investigation of the number 342.

Three hundred and forty-two   by Laura Stuart

I used my calculator to divide 342 by all the numbers from 1 to 20.

If I got decimals on my calculator I knew it didn't divide exactly.

Here I've written out the first few.

DIVIDING NUMBER	CALCULATOR'S ANSWER	MY COMMENT
1	342	Exact
2	121	Exact
3	114	Exact
4	85·5	Not Exact
5	68·4	Not Exact

When I had finished up to 20 I counted all the exact ones. I got 7 exact numbers and 13 not exact. 7 and 13 come to 20 so I know I counted them all.

My friend Jennifer chose the number 345 and she only got 4 exact. So my number beat hers by 3.

My teacher said that there are numbers less than a 1000 that would beat my number. I hope nobody finds one because my number is the best in the class so far.

I'm still trying to find a better number but so far I've had no luck.

1 Choose your own number and write a report on it.

2 'Why 360?' Write a report on 360. Does choosing an angle that fits around a point 360 times to measure other angles seem like a good idea to you? Explain your answer.

Handy angles

1 Place your hand on a blank sheet of paper. Spread your fingers as wide as you can. Now trace round the outline of your hand.

You will find that the directions of your fingers can be extended back to meet at a point.

2 Copy this table, and use a protractor to complete it.

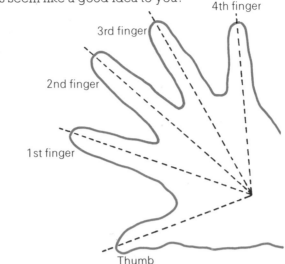

4th finger

3rd finger

2nd finger

1st finger

Thumb

MY FINGER CHART

ANGLE BETWEEN → ↓	4th finger	3rd finger	2nd finger	1st finger
Thumb				
1st finger				
2nd finger			
3rd finger			

3 a Which measurement should be used for the 'spread' of the hand?

b Who has the largest hand-spread in the class?

c Do small hands give small angles and large hands large angles?

d What sort of measurements would piano players like for their hands? Why?

DRAWING ANGLES

A flowchart

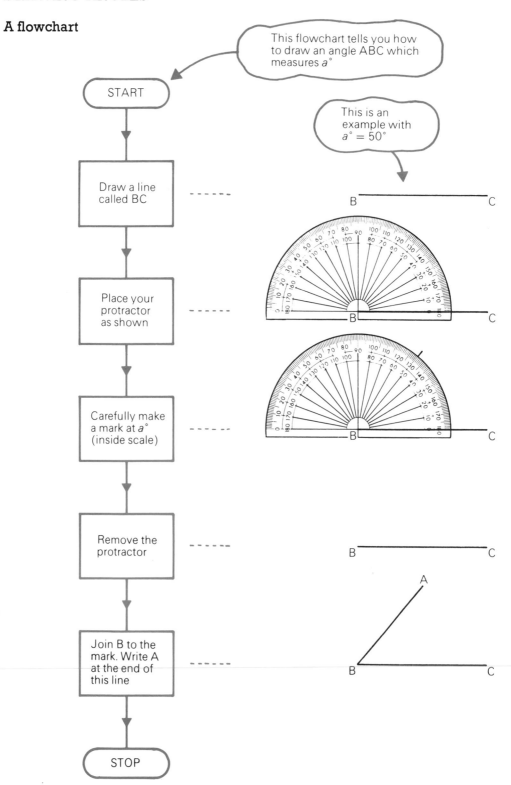

This flowchart tells you how to draw an angle ABC which measures $a°$

START

This is an example with $a° = 50°$

Draw a line called BC

B ———————————— C

Place your protractor as shown

Carefully make a mark at $a°$ (inside scale)

Remove the protractor

Join B to the mark. Write A at the end of this line

STOP

1A Use the flowchart to draw an $\angle ABC$ with $a° = 60°$.

2A Now try the flowchart with these values for $a°$:
   a 45°   **b** 130°   **c** 98°   **d** 90°   **e** 9°

3A Draw these angles using the outside scale.
   They should all face the other way.
   a $\angle PQR = 35°$
   b $\angle GHS = 100°$
   c $\angle UVW = 150°$

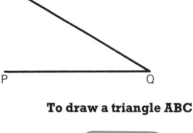

4B Use this flowchart with
$T = 10$, $x = 50$ and $y = 30$.
Measure $\angle BAC$ and call it $z°$.
Calculate $x° + y° + z°$.

To draw a triangle ABC

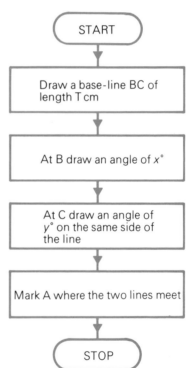

START

Draw a base-line BC of length T cm

At B draw an angle of $x°$

At C draw an angle of $y°$ on the same side of the line

Mark A where the two lines meet

STOP

5B Copy and complete this table. Use a base-line 6 cm long.

$x°$	$y°$	$z°$ (measure)	$x° + y° + z°$
50°	30°		
60°	40°		
120°	20°		
94°	47°		

Write about your results, describing anything you have found out.

Can you draw the angle?

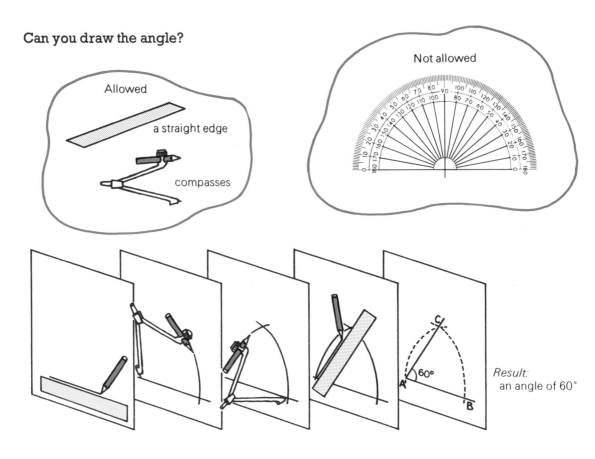

Allowed

a straight edge

compasses

Not allowed

Result:
an angle of 60°

=== *Exercise 8* ===

Copy these diagrams, and draw the angles.

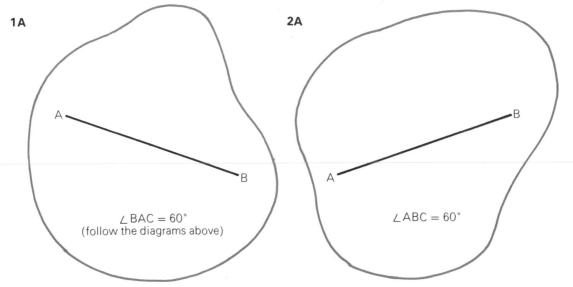

1A

∠ BAC = 60°
(follow the diagrams above)

2A

∠ ABC = 60°

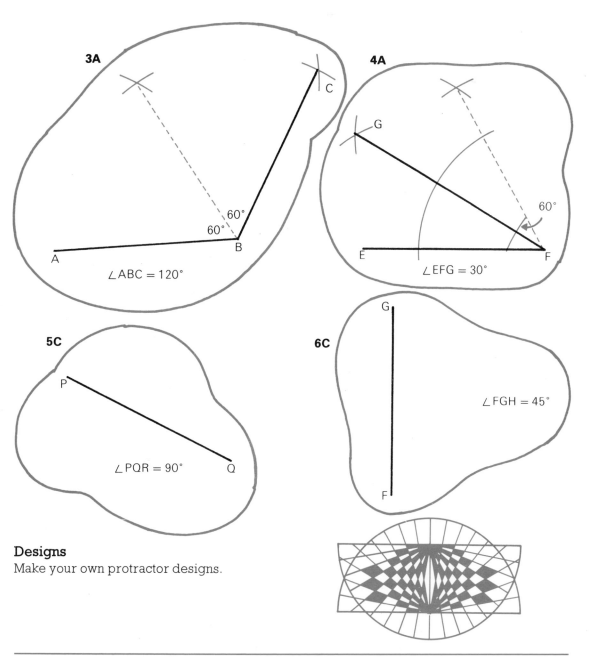

3A

A ———————— B

∠ABC = 120°

60°
60°
C

4A

G

60°

E ———————— F

∠EFG = 30°

5C

P

∠PQR = 90°

Q

6C

G

∠FGH = 45°

F

Designs

Make your own protractor designs.

CHECK-UP ON **INTRODUCING ANGLES**

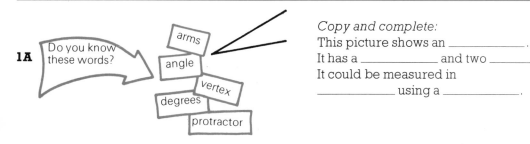

1A

Do you know these words?

arms
angle
vertex
degrees
protractor

Copy and complete:
This picture shows an _____.
It has a _____ and two _____.
It could be measured in
_____ using a _____.

2A

Can you name an angle?

Can you measure an angle in degrees?

Can you do these?

Can you draw a given angle?

a Name three angles in this drawing.

b Measure this angle.

c ∠ABC = 65°. Draw this angle.

3A, B

Can you solve these problems?

a How many 30° angles are there in this diagram?
b Name them all.

c What is the connection between *a* and *b* in these tiles?

4C

Investigate

a

b

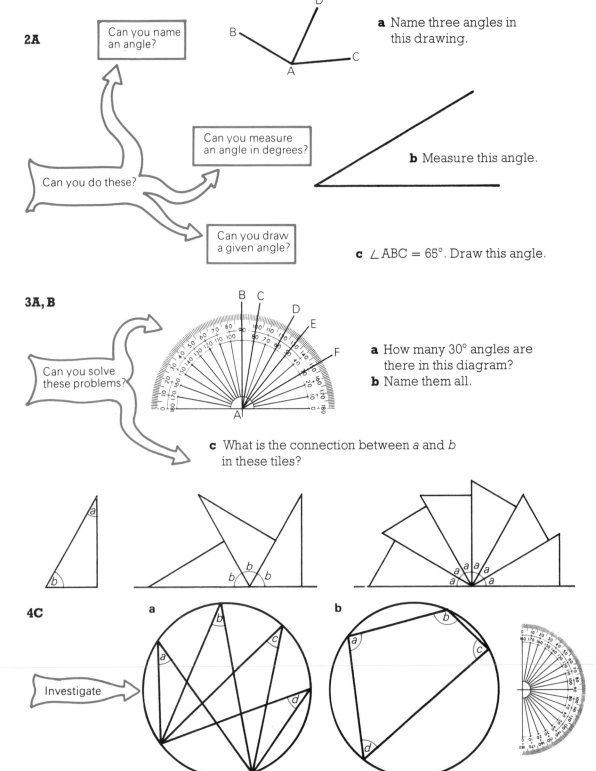

COUNTING IN TENS

Class discussion

1 Television programmes often show the date when they were made in Roman numerals.
Can you work out the date on this television screen? Why is it difficult to do this?

2 Our decimal system is based on the ten numerals 0, 1, 2, 3, 4, 5, 6, 7, 8, 9 and on counting in tens. Why do you think ten is used?
This is called a place system, because each place has ten times the value of the one on its right.

Th	H	T	U	
			1	× 10
		1	0	× 10
	1	0	0	× 10
1	0	0	0	

By changing its place, 1 can stand for one unit, or one ten, or one hundred, . . .

Read out these numbers to show their meaning:

555   3164   90   22   1070   25 348

3 Look at the athletics scoreboard in the picture.
The decimal point separates the whole numbers from the fractions.

H	T	U	t	h
1	0	6	8	2

What do **t** and **h** stand for?

Read out these numbers to show their meaning:

4·5   7·23   36·4   20·6   0·3   11·11   0·09

Note: Where there is no whole number we put a zero before the decimal point.
Why do you think this is done?

4 If you think of the meaning, you can write a decimal in the form of a fraction. 0·5 means 5 tenths, or $\frac{5}{10}$.
Read these out as decimals and as fractions:

0·1   0·9   0·6   0·4   0·7

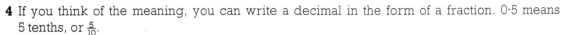

5 0·24 means 2 tenths and 4 hundredths, or 24 hundredths, or $\frac{24}{100}$. Read these out as decimals and as fractions:

0·31   0·17   0·50   0·99   0·03

6 When you make out a cheque you have to write the sum of money in words and in figures. Write out these sums of money in words:

£349   £2071   £5·43   £78·09   £1234·98

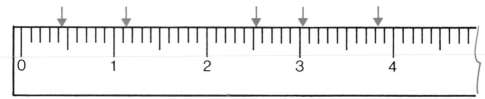

=================== *Exercise 1A* ===================

1 Write down the next three numbers to the right, and the next three numbers to the left, in the sequence . . . 100, 10, 1, . . .

2 a Use the numerals 0, 1, 2, . . . , 9 to write down these numbers:

Fifty-seven   Two hundred and forty-six   Eighty-eight

One hundred and one   Two thousand five hundred and thirty-nine

b Underline the tens figure in each one.
c Circle the units figure in each one.

3 a Write down these numbers in figures:

Five point six   Thirty-one point two   Nought point three

Six hundred and forty-one point two   One hundred and four point nought nine

b Underline the tenths figure in each one.

4 Write these numbers in decimal form:

H	T	U	t	h
	5	3	1	7
		2	3	4
	2	0	3	5
	5	1	0	6
3	4	0	1	
2	0	0	1	3
			2	7
				6

5 What numbers on this ruler are the arrows pointing to?

6 Draw or trace a ruler like the one above. Mark arrows at:
  a 3·5   **b** 1·7   **c** 2·9   **d** 0·1   **e** 0·9

7 a In the number 25 348, what is the value of: (i) the 5   (ii) the 4?
  b In the number 3·69, what is the value of: (i) the 3   (ii) the 9?

8 Which number in each pair is greater?

  a 13 406 or 13 460   **b** 1·9 m or 2·1 m   **c** £10·01 or £9·99

  d 76·54 or 76·45   **e** 3·09 cm or 3·10 cm   **f** 0·02 or 0·1

9 Use your calculator carefully to find the answers to the following in decimal currency form:

  a £1 + 5p   **b** £1 − 9p   **c** £3 ÷ 2   **d** £1·01 × 10

10 a What numbers on this timer are the arrows pointing to?

  b Arrange the numbers in order, from smallest to largest.

11 Write these numbers as fractions (or as mixed numbers, like $2\frac{3}{10}$):

  a 3·7   **b** 5·1   **c** 6·9   **d** 0·3   **e** 11·1

  f 0·1   **g** 0·13   **h** 0·99   **i** 1·07   **j** 30·03

12 Write these numbers in decimal form.

  a $1\frac{3}{10}$ **b** $1\frac{4}{10}$ **c** $\frac{7}{10}$ **d** $\frac{1}{10}$ **e** $2\frac{13}{100}$ **f** $8\frac{47}{100}$

  g $\frac{23}{100}$ **h** $4\frac{9}{10}$ **i** $\frac{9}{100}$ **j** $\frac{6}{10}$ **k** $\frac{6}{100}$ **l** $209\frac{1}{100}$

13 Arrange these numbers in order, from smallest to largest:

14·78   14·5   15·01   14·99   14·28   14·93   14·75

===== *Exercise 1B* =====

1 In the number 1 234 567·890, what is the value of:

  a the 1   **b** the 3   **c** the 5   **d** the 7   **e** the 9   **f** the 0?

2 Arrange these numbers in order, from smallest to largest:

  10·01, 9·95, 10·104, 10·004, 9·876, 10

3 Write in decimal form: **a** $1\frac{9}{10}$ **b** $1\frac{9}{100}$ **c** $1\frac{9}{1000}$.

  Explain how the zeros change the place value of the 9 in your answers.

4 This ruler contains a magnified section between 4 and 5. What numbers are the seven arrows pointing to?

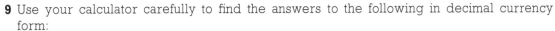

5 Maureen had just passed her driving test, and was checking the pressure in her tyres. What is the reading on the gauge? (**Careful!** Each division is not 0·1).

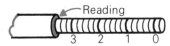
Reading

6 What are the readings on these gauges?

  a   **b**

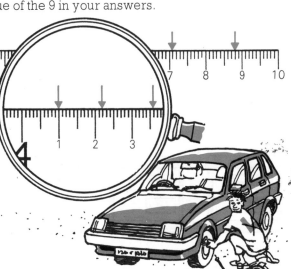

The binary system is a place system based on two numerals, 0 and 1, and counting in twos. Compare the decimal system on page 119, question **2**.

1 What numbers would these place headings stand for? **S T S E F T U**.

2 How would the numbers 1, 2, 3, 4, 5, 6, 7, 8, 10, 12, 20, 40 and 100 be listed under these headings as binary numbers?

3 List some more binary numbers, and then write down the decimal form of each. You might like to find out more about the binary system and its influence on the development of computers.

ADDITION AND SUBTRACTION

$$
\begin{array}{r}
£\ 4·75 \\
3·30 \\
1·96 \\
\hline
£10·01 \\
\end{array}
\qquad
\begin{array}{r}
3·44 \\
-1·26 \\
\hline
2·18 \\
\end{array}
$$

Our British currency is based on the decimal number system. You know that when you are adding and subtracting sums of money the decimal points must be kept in line, one below the other. Why is this?

Calculator left at home! Try this exercise without it.

1 Copy and complete:

a 3·56 +2·16	**b** 8·43 +9·65	**c** 12·35 +8·15	**d** 15·29 +16·44	**e** 123·8 +207·5
f 8·65 −2·61	**g** 9·83 −1·16	**h** 10·44 −6·62	**i** 12·8 −6·9	**j** 234·5 −161·5

2 Add:
   a 8·7 and 7·8   **b** 3·9 and 9·3   **c** 6 and 8·8   **d** 21·5 and 16·6
   e 18 and 2·12   **f** 25·5 and 3·8   **g** 6·8 and 11·4   **h** 23·8 and 1·19

3 Subtract:
   a 6·5 from 9·2   **b** 6·8 from 8·6   **c** 15·3 from 20·9   **d** 27·8 from 30
   e 56·7 from 88   **f** 2·14 from 7·5   **g** 3·5 from 6·15   **h** 1·19 from 23·8

4 Here are the readings on four petrol pumps.

| £ 20·00 | £ 18·00 | £ 10·00 | £ 22·00 |
| 35·5 litres | 32·9 litres | 17·7 litres | 39·4 litres |

   a How much did the four motorists spend on petrol altogether?
   b What was the total amount of petrol put in their tanks?

DECIMALS IN ACTION

5 David Murray was given £10 for his birthday. He bought a paperback book for £1·95, a knife for £3·16 and a box of chocolates for £1·54. He paid for them with his £10 note and was given £3·25 change. Was this the correct change?

6 Calculate the perimeters of these shapes:

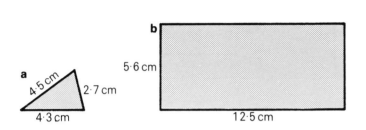
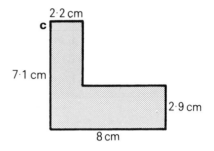

7 In May 1954, Roger Bannister became the first person to run a mile in less than four minutes. He ran the four laps of the race in 57·5, 60·7, 62·3 and 58·9 seconds.

   a What was his total time?   **b** By how much did he break the four minute barrier?

8 In a diving competition, Chris scored 43·56, 37·27 and 53·26 points. Gary scored 51·26, 33·76 and 49·37. Who won? By how many points?

═══════════════════════ *Exercise 2B* ═══════════════════════

Again, no calculator!

1 Add:
  a 2·8 and 3·7   **b** 5·63 and 2·19   **c** 18 and 12·6
  d 87·6 and 8·76   **e** 165·3 and 14·08   **f** 23·56 and 27·956

2 Subtract:
  a 8·8 from 100   **b** 1·201 from 10   **c** 235·9 from 1000

3 Calculate:
  a 2·35+11·5−9·9   **b** 14·3−8·56+2·75   **c** 29·3−5·65−0·06

4 The dimensions of these cars are shown in metres.

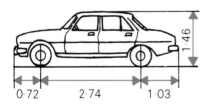
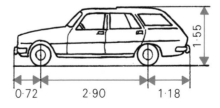

Calculate: **a** the difference in the lengths of these two cars;
          b the difference in their heights.
The wheelbase is the distance between the front and back wheels.
What is the difference between the wheelbases of the two cars?

5 Jacqui Green buys a cassette recorder for £27·50, a pack of cassettes for £3·35 and a small transistor radio for £15·09. What change will she be given from a £50 note?

6 Calculate the perimeters of these shapes:

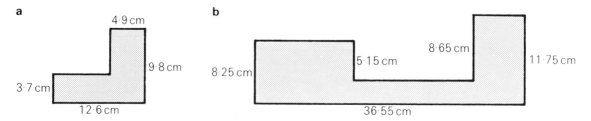

a

4·9 cm

9·8 cm

3·7 cm

12·6 cm

b

8·65 cm

5·15 cm

8·25 cm

11·75 cm

36·55 cm

7 An electrician has 100 metres of wire on a drum. He uses 13·6 metres and 28·7 metres of wire in one house, and the same lengths again in another house. What length of wire will be left?

8 In an indoor athletics stadium four laps of the track are run in an 800 metre race. An athlete ran the first lap in 24·9 seconds, the first two laps in 51·4 seconds and the first three laps in 78·9 seconds. Her time for 800 metres was 1 minute 46·5 seconds.
  a What were her times for the second, third and fourth laps?
  b Which was her fastest lap?

Exercise 2C

World ice dance championship
Here are the marks for the winning pair in the compulsory dances.

	Judges								
	1	2	3	4	5	6	7	8	9
Quickstep	5·8	5·9	5·7	5·8	5·8	5·8	5·7	5·8	5·9
Waltz	5·9	5·9	5·8	5·9	5·9	5·9	5·8	5·9	5·8
Tango	5·9	5·9	5·9	5·9	5·9	5·9	5·9	5·9	5·9

1 Calculate the total points awarded by each of the nine judges for the three dances.

2 Calculate the nine judges' total scores for:
  a the quickstep　　　　**b** the waltz　　　　**c** the tango.

3 a Add together the total points in question **1**.
  b Add together the total points in question **2**.
  c Are your totals in **a** and **b** the same?

4 The final points awarded in the championship were:

　　　　　　Pair A: 5·906　　　　**Pair B**: 5·656　　　　**Pair C**: 5·692

  a Who came second?
  b What were the differences in scores between first and second, and between second and third?

5 Alison's father is an architect. Here is the ground plan of a house that he is designing. The lengths are in metres.
Calculate the length of:

a the east wall **b** the south wall.

He intends to lay a path of square $\frac{1}{2}$ metre paving slabs right round the outside walls of the house. Make a rough sketch, and estimate the number of paving slabs that would be needed.

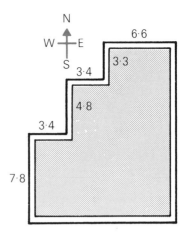

MULTIPLICATION BY A WHOLE NUMBER

Times 10

In the chapter on **Numbers in Action** you saw that when a number is multiplied by 10, each figure moves 1 place to the left.

H	T	U	
	1	5	× 10
1	5	0	

$15 \times 10 = 150$

H	T	U	t	
		8	5	× 10
	8	5	0	

$8.5 \times 10 = 85.0$, or 85

A simple rule is:
To multiply by 10, move the decimal point 1 place to the right.
To multiply by 100, move the decimal point 2 places to the right, and so on.

Examples

$$1.7 \times 10 = 17 \qquad\qquad 5.83 \times 100 = 583$$
$$3.14 \times 10 = 31.4 \qquad\qquad 2.7 \times 100 = 270$$

Times other whole numbers

When multiplying a decimal by a whole number you must be careful to put the decimal point in the correct place in the answer.

 It is a good idea to make a rough estimate of the answer.

A rough estimate for 7.43×18 is $7 \times 20 = 140$.

$$7.43 \times 18 = 133.74.$$

133·74 is a sensible answer. 13·374, or 1337·4 would not be.
Notice that the number of figures after the decimal point is the same before and after the multiplication.

DECIMALS IN ACTION

Try questions **1–3** without using a calculator.

1 Multiply each of these by 10:
  a 3·4   **b** 4·7   **c** 15·8   **d** 251·9   **e** 1·35   **f** 0·68

2 Multiply each of these by 100:
  a 4·56   **b** 2·13   **c** 7·8   **d** 8·91   **e** 0·52   **f** 0·063

3 In each of these, write down a rough estimate for the answer. Then calculate the exact answer.
  a 6·3 × 5   **b** 8·2 × 7   **c** 9·4 × 8   **d** 7·8 × 5
  e 12·4 × 9   **f** 15·6 × 4   **g** 23·4 × 3   **h** 35·2 × 6

4 Which is the most sensible answer in each of the following?

  a Lila's bike was travelling at: (i) 0·05 mph (ii) 5 mph (iii) 500 mph.

  b The average height of pupils in the class is: (i) 1·7 m (ii) 0·17 m (iii) 17 m.

  c The cost of a strawberry tart is: (i) 0·38p (ii) 380p (iii) 38p.

  d John's mark out of 20 in a test is: (i) 150 (ii) 0·15 (iii) 15.

5 Douglas knows a friendly shopkeeper who allows him to pay a weekly sum of money for a personal stereo. If the charge is £1·45 a week for 15 weeks, how much does Douglas pay?

6 Helen is collecting woollen rags to raise money for charity. She collects 15·6 kg of woollen rags and the rag merchant pays 18p per kg. How much money does she receive?

7 Calculate Mr Williams' telephone bill. Start by finding how many units he used between the readings on 18 March and 16 June.

The Talky Telephone Co. BILL	
RENTAL FOR 3 MONTHS	£ 17·75

DATE	READING
18 MARCH	5708
16 JUNE	6066

.... UNITS at 5·25p = £ _____

TOTAL £ _____

8 Which method of buying is cheaper? By how much?

£168·95 cash OR
£45 deposit and
26 weekly
payments of £4·99

PORTABLE TV

9 Avril has a part-time job in a supermarket. She earns £18·35 a week, which is free of tax. After saving her earnings for 3 weeks, she buys a radio cassette recorder costing £47·98. How much money will she have left?

10 A builder needs steel rods 8·5 metres long to make reinforced concrete pillars. He has to order 175 rods. What is the total length of steel rod required?

=================== *Exercise 3B* ===================

1 Multiply each of these by 10, 100 and 1000:
 a 3·92 **b** 0·15 **c** 22·06 **d** 1·032 **e** 0·004

2 Write down an approximation for each of these. Then calculate the exact answer.
 a 2·4 × 38 **b** 7·2 × 13 **c** 29·6 × 19 **d** 57·79 × 45

3 Christine goes on holiday to France. She takes £156 with her, which she exchanges for French francs. If she is given 11·79 francs for each £1, how many francs will she receive altogether?

4 A machine sends out tennis balls so that a player can practise his return shots. It delivers one ball every 12·8 seconds. If the machine holds 50 balls, how long does it take to empty?

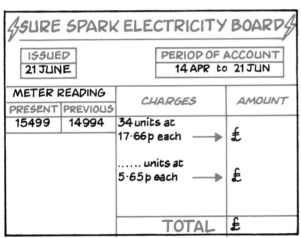

5 Calculate the total cost of electricity charged in this bill, by first finding:
 a the number of units used
 b the number of units at 5·65p each
 c the cost of units at 17·66p, and at 5·65p.

6 Julie and her friend Susan decide to save for a holiday. Julie saves £4·75 every month and Susan saves £1·25 every week. How much will each girl have saved after a year?

7 Derek delivers papers and is paid £12·50 a week. Each week he buys a magazine costing 75p and he gives his younger sister 35p a week. He saves the rest for 6 weeks to buy a camera costing £59·95.
If rolls of film cost £3·42 each, how many can he buy with the money he has left?

MULTIPLICATION BY A DECIMAL

Again you must be careful to put the decimal point in the correct place in the answer.

DECIMALS IN ACTION

================= *Exercise 4* =================

Calculate:

1 a 36×37 **b** $3 \cdot 6 \times 37$ **c** $3 \cdot 6 \times 3 \cdot 7$ **d** $0 \cdot 36 \times 3 \cdot 7$

2 a $4 \cdot 3 \times 6 \cdot 8$ **b** $0 \cdot 43 \times 6 \cdot 8$ **c** $0 \cdot 43 \times 0 \cdot 68$ **d** $0 \cdot 043 \times 0 \cdot 68$

3 For questions **1** and **2**:

   a Count the number of figures after the decimal point in each answer.

   b Count the total number of figures after the decimal point in the two numbers being multiplied.

   What is the connection between these results?

$4 \cdot 3 \times 6 \cdot 8$	$4 \cdot \underline{3}$	1 figure after the decimal point
↑ ↑	$\times 6 \cdot \underline{8}$	1 figure after the decimal point
tenths × tenths = hundredths.	3 4 4	
So the answer must have a	25 8 0	
hundredths figure.		
	$29 \cdot \underline{2\,4}$	2 figures after the decimal point

The rule for finding the position of the decimal point is:

Count the total number of figures after the decimal point in the two numbers being multiplied. This gives the number of figures after the decimal point in the answer.

================= *Exercise 5A* =================

1 Before calculating the following, write down the number of figures after the decimal point in the answer, as a check.

   a $3 \cdot 5 \times 1 \cdot 8$ **b** $0 \cdot 67 \times 5 \cdot 4$ **c** $45 \cdot 2 \times 0 \cdot 9$ **d** $5 \cdot 3 \times 0 \cdot 026$

2 Linda needed 8·5 metres of curtain material. It cost £9·99 a metre. How much did she have to pay?

3 1 kilogram of beef costs £7·42. How much would 2·35 kg cost?

4 Value Added Tax (VAT) is added to the cost of things you buy. The amount of VAT is found by multiplying the cost by 0·15. Calculate the VAT charged on these items:

a

TV Set

b

Fridge

c

Camera

5 Mr Smart fills up the tank of his car with 44·3 litres of petrol. 1 litre costs 56·4 pence. How much will he be charged?

6 Calculate the cost of this gas account.

THE DODGY GAS COMPANY		ACCOUNT
DATE OF READING	GAS SUPPLIED	CHARGES
13 JUNE	340·734 units at 44·8p per unit ⟶	£
	STANDING CHARGE	£ 11·45
	TOTAL	£

===== *Exercise 5B* =====

In a diving contest each competitor's final score was calculated like this:
The three judges' scores were added together, and then multiplied by the 'degree of difficulty' number for the dive.

1 Elaine was awarded 5, 5·5 and 5·5 for a dive with a degree of difficulty of 1·4. What was her final score?

2 Alan had scores of 6, 6·5 and 5·5 for a dive with a degree of difficulty of 1·8. Find his final score.

3 Who won this four-dive competition? Arrange your answers in a table.

	Hari	Laura	Craig
First Dive	6, 6·5, 5: 1·8	7, 5·5, 5·5: 1·6	5·5, 5, 6: 2·1
Second Dive	4·5, 4, 3·5: 2·4	6·5, 6·5, 7: 1·4	3·5, 3, 4: 2·2
Third Dive	5·5, 5, 4: 1·9	6·5, 6, 7: 1·6	4, 4·5, 3: 2·1
Fourth Dive	6, 6·5, 6: 1·8	7·5, 7, 6: 1·5	6·5, 6, 5: 2·3

4 A self-employed tradesman completes a job in 8·5 hours. He charges £8·75 an hour. How much did the job cost?
How much does he earn in a 39½ hour week, before deductions for income tax, etc?

5 By mistake, the cost of 19·5 metres of material at £6·28 per metre was calculated by multiplying £62·8 by 1·95. Explain what has happened to the result.

6 Can you find short-cuts for doing these calculations mentally?
  a Add 6·7, 6·3, 4·5, 5·5, 5·6, 2·4, 8·1, 8·9.
  b Multiply £9·99 by 9.
  c Find the cost of 18 stamps at 17p and 18 at 13p.
  d How much would you pay for six shirts costing £14·75 each and six ties each costing £5·25?

DIVISION BY A WHOLE NUMBER

Class discussion—Rounding off a decimal

Divide 23 by 19.
The calculator shows 1·2105263.
This is the answer to 7 decimal places.
We do not often need such an accurate answer, so we 'round it off' to a suitable number of decimal places.

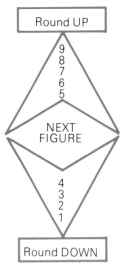

Round UP

9
8
7
6
5

NEXT
FIGURE

4
3
2
1

Round DOWN

$$1·2105263 = 1·210526,\text{ rounded off to 6 decimal places}$$
$$= 1·21053, \quad,, \quad,, \quad,, \quad 5 \quad,, \quad,,$$
$$= 1·2105, \quad,, \quad,, \quad,, \quad 4 \quad,, \quad,,$$
$$= 1·211, \quad,, \quad,, \quad,, \quad 3 \quad,, \quad,,$$
$$= 1·21, \quad,, \quad,, \quad,, \quad 2 \quad,, \quad,,$$
$$= 1·2, \quad,, \quad,, \quad,, \quad 1 \quad,, \quad,,$$
$$= 1, \quad,, \quad,, \quad,, \quad\text{the nearest whole number}$$

Discuss the rules that are used for rounding off to a number of decimal places.

DECIMALS IN ACTION

========== *Exercise 6* ==========

1 Round these off to 1 decimal place:
  a 5·18 **b** 6·33 **c** 7·24 **d** 8·77 **e** 3·15 **f** 4·09
  g 1·55 **h** 12·46 **i** 15·02 **j** 1·234 **k** 2·461 **l** 3·157

2 Round these off to 2 decimal places:
  a 3·147 **b** 2·013 **c** 6·875 **d** 0·194 **e** 0·035 **f** 10·246

3 Round these off to the nearest penny:
  a £5·632 **b** £0·285 **c** £1·914 **d** £12·117

4 In these calculations, round off your answers to 2 decimal places:
  a 28 ÷ 13 **b** 45 ÷ 23 **c** 7·4 ÷ 17 **d** 16·1 ÷ 11 **e** 1 ÷ 7

5 These measurements do not have to be exact. Round them off:
  a 36·8 mm to the nearest millimetre   **b** 2·54 cm to the nearest tenth of a centimetre
  c 8·27 g to the nearest gram   **d** 2·351 km to the nearest tenth of a kilometre
  e 16·324° to the nearest hundredth of a degree.

6 What is the 'best' answer for £5 ÷ 3?

7 Round off 3·72468 to 1, 2, 3 and 4 decimal places.

8 Round off 29·98055 to 1, 2, 3 and 4 decimal places.

Division

 It is a good idea to make a rough estimate of the answer.

A rough estimate for $26 \cdot 56 \div 8$ is $24 \div 8 = 3$.
$26 \cdot 56 \div 8 = 3 \cdot 32$.
$3 \cdot 32$ is a sensible answer. $0 \cdot 332$ or $33 \cdot 2$ would not be.

=========================== *Exercise 7A* ===========================

1 In each of these write down a rough estimate. Then calculate the exact answer.
   a $29 \cdot 64 \div 3$ **b** $9 \cdot 645 \div 5$ **c** $7 \cdot 384 \div 8$ **d** $10 \cdot 86 \div 6$
   e $15 \cdot 87 \div 4$ **f** $123 \cdot 21 \div 3$ **g** $93 \cdot 38 \div 9$ **h** $1 \cdot 869 \div 2$

2 Eight friends order ice-cream. The bill comes to £2·24. How much should each pay?

3 Calculate, to the nearest penny where necessary:
   a £7·65 ÷ 9   **b** £157·84 ÷ 12   **c** £13·62 ÷ 7   **d** £1·75 ÷ 14

4 Carol buys 7 balls of wool for £5·99. She uses only 6 of them and sells the seventh to her friend Lesley. How much should Carol ask her to pay?

5 A tin of Miaow cat food costs 38p. A shop advertises that if a customer buys five tins, the cost is £1·75. How much cheaper does this make each tin?

6 a The distance between two towns by rail is 31 kilometres. A day return ticket costs £5·60. What is the cost in pence per kilometre (to 2 decimal places)?
   b The distance between two different towns by rail is 43 kilometres, and a day return costs £5·80. What is the cost in pence per kilometre (to 2 decimal places) this time?
   c Which is better value and by how much?

7 How many strips of metal 15·7 cm long can be cut from a strip 250 cm long? What length is left over (to the nearest tenth of a cm)?

8 A racing car went round the track in times of 61·3, 58·7, 59·2, 57·9 and 56·8 seconds. Add these times and then divide by 5 to find the average lap time (to the nearest tenth of a second).

=========================== *Exercise 7B* ===========================

1 Write down an approximation for each of these. Then calculate the answer to 3 decimal places, where necessary.
   a $246 \cdot 3 \div 8$ **b** $90 \cdot 01 \div 12$ **c** $34 \cdot 5 \div 15$ **d** $84 \cdot 26 \div 32$

2 a Is $23 \times 19 = 19 \times 23$?   **b** Is $23 \div 19 = 19 \div 23$?   When does the order matter?

3 A hockey team manager orders 11 hockey skirts for £95·24. How much will each girl have to pay for her skirt?

4 A school football team sends for 15 jerseys and pays a total of £88·75 for them. The first day they are used one member loses his and is told that he must pay to replace it. How much must he pay?

5 Members of a youth club decided to try to collect a kilometre of pennies. A kilometre is 100 000 centimetres, and 1 penny is 2 centimetres across.
How many coins would they have to collect? How much would these be worth?

6 Bert is a worker on a building site. A load of sand weighing 4086 kg is tipped out of a lorry. Bert has to take this sand in barrow-loads to a mixing machine. It takes 162 barrow-loads to remove all the sand, and Bert carries the same amount of sand each time. What is the weight of sand in each barrow-load? Round off your answer to 1 decimal place.

7 Margaret thinks that she can calculate the thickness of the page of a book from these measurements.
Calculate the thickness in millimetres, correct to 2 decimal places.
Try to do this with one of your own books.

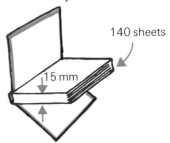

140 sheets

15 mm

8 A supermarket sells two sizes of the same washing-up liquid. The 450 ml size costs 42p and the 750 ml size costs 65p. Which is the better buy? Why?

FRACTIONS AND DECIMALS

$\frac{2}{3} = 2 \div 3 = 0·66666\ldots$, a 'recurring decimal'.
It is usual to round this off to 2 or 3 decimal places.
$\frac{2}{3} = 0·667$, rounded off to 3 decimal places.

───────── *Exercise 8* ─────────

Use division to express each of these fractions in decimal form.
Where necessary, round off your answers to 3 decimal places.

1 $\frac{1}{5}$	**2** $\frac{2}{5}$	**3** $\frac{3}{5}$	**4** $\frac{4}{5}$	**5** $\frac{1}{4}$	**6** $\frac{1}{2}$	**7** $\frac{3}{4}$
8 $\frac{1}{8}$	**9** $\frac{3}{8}$	**10** $\frac{5}{8}$	**11** $\frac{7}{8}$	**12** $\frac{1}{3}$	**13** $\frac{5}{6}$	**14** $\frac{1}{7}$
15 $\frac{2}{7}$	**16** $\frac{3}{7}$	**17** $\frac{5}{9}$	**18** $\frac{7}{11}$	**19** $\frac{9}{13}$	**20** $\frac{1}{101}$	**21** $\frac{99}{100}$

22 Copy and complete this box of useful results.

$\frac{1}{4} =$	$\frac{1}{3} =$	$\frac{1}{10} =$
$\frac{1}{2} =$	$\frac{2}{3} =$	$\frac{1}{5} =$
$\frac{3}{4} =$		$\frac{1}{8} =$

23 Try to find fractions which are equal to: **a** $0·1666\ldots$ **b** $0·0909\ldots$ **c** $0·4444\ldots$

DIVISION BY A DECIMAL

If a calculator is not available, you can divide the numbers like this. Multiply numerator and denominator by 10, 100, ... in order to have a whole number in the denominator. Then divide by the whole number.

$$14 \cdot 84 \div 0 \cdot 7$$

$$= \frac{14 \cdot 84 \times 10}{0 \cdot 7 \times 10} \quad \text{(numerator)} \atop \text{(denominator)}$$

$$= \frac{148 \cdot 4}{7}$$

$$= 21 \cdot 2$$

========== Exercise 9 ==========

1 No calculator in this one! Calculate:
  a $4 \cdot 2 \div 0 \cdot 6$ **b** $9 \cdot 76 \div 0 \cdot 4$ **c** $10 \cdot 5 \div 0 \cdot 5$ **d** $84 \cdot 6 \div 0 \cdot 2$
  e $3 \cdot 88 \div 0 \cdot 02$ **f** $9 \cdot 66 \div 0 \cdot 07$ **g** $0 \cdot 615 \div 0 \cdot 03$ **h** $0 \cdot 126 \div 0 \cdot 9$

2 Round off the answers to the following to 3 decimal places:
  a $2 \cdot 843 \div 0 \cdot 71$ **b** $2 \cdot 683 \div 4 \cdot 5$ **c** $2 \cdot 596 \div 0 \cdot 08$ **d** $90 \cdot 63 \div 0 \cdot 066$

3 a Divide 5·82 by 0·06.   **b** Divide 36·54 by 2·9.

4 A girl bought 3·5 metres of dress material and paid £24·15 for it. How much per metre was the shop charging for the material?

5 The organisers of a Youth Club have calculated the total cost of running a Disco to be £210. If tickets are sold at £1·75 each, how many must be sold to cover the cost?

6 A butcher charges £5·45 for a joint of meat. If the price is £4·36 per kilogram, what is the weight of the joint of meat?

7 The total score for a dive in a diving competition is found by adding the three judges' scores together and multiplying by the degree of difficulty of the dive. Copy and complete this table:

Diver	Judges' Scores	Degree of Difficulty	Total score
Sarah	6·5, 6, 6	1·7	
Nadia	7, 6·5, 7·5	1·9	
Cathy	7, 7, 6·5		32·80
Jasmine	6, 7·5, 6		40·95

1 A walker reckons he can walk 4·25 miles in an hour. How far can he walk in $3\frac{1}{2}$ hours if he keeps up this steady pace?

2 How many car parking spaces 4·8 metres wide can be fitted into a street 123·4 metres long?

3 A secretary works a 37 hour week for which she is paid £5·52 per hour. What is her weekly wage? How much does she earn in a year?

4 It is estimated that it will cost a school £275 to put on an end of term concert. If tickets for the concert are sold at £1·50 each, how many must be sold to cover the cost?

5 A joiner took 32·5 hours to build a garden fence and was paid £217 for his work. He returned later for 1 hour to spray the fence with creosote. How much should he be paid for this hour's work?

6 The three front row forwards in Puffin High School's rugby team weigh 76·4 kg, 68·7 kg and 75·8 kg. The three front row forwards in Blowmore Academy's rugby team weigh 77·3 kg, 66·4 kg and 76·1 kg. Which team has the heavier front row and by how much?

7 A gardener wants to buy grass seed. One packet of seed holds 485 grams and costs £1·36. A packet from a different supplier contains 510·5 grams and costs £1·48. Which is the better buy?

8 Fence posts have to be put round the sides of a square playground. Each side is 26·25 metres long, and the posts have to be 3·75 metres apart. How many posts are needed? Show this in a sketch.

PUZZLES

1 Copy and complete this cross-number puzzle:

Clues

Across
A 153·4 + 415·6
E 43·21 × 100
G 13·5 ÷ 0·9
H 13·2 × 5
I 976·3 + 814·7
K 450 × 0·9

Down
B 7543·2 − 1086·2
C 65·1 ÷ 0·7
D 0·6 × 6 × 60
F 8700 × 0·3
G 76·7 + 87·5 − 47·2
J 37·6 ÷ 0·4

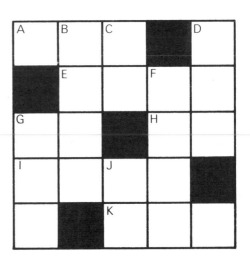

2 Here is a code:

A	B	C	D	E	F	G	H	I	J	K	L	M
↑	↑	↑	↑	↑	↑	↑	↑	↑	↑	↑	↑	↑
0	0·1	0·2	0·3	0·4	0·5	0·6	0·7	0·8	0·9	1	1·1	1·2

N	O	P	Q	R	S	T	U	V	W	X	Y	Z
↑	↑	↑	↑	↑	↑	↑	↑	↑	↑	↑	↑	↑
1·3	1·4	1·5	1·6	1·7	1·8	1·9	2	2·1	2·2	2·3	2·4	2·5

Find the message:

$\|\, 2 \times 0{\cdot}9 \,|\, \frac{1}{5}$ as a decimal $|\, 4{\cdot}2 \div 6 \,|\, 0{\cdot}2 \times 7 \,|\, 12{\cdot}6 \div 9 \,|\, 1{\cdot}21 \div 1{\cdot}1 \,\|\, 0{\cdot}23 + 0{\cdot}47 \,|$

$|\, 23{\cdot}8 \div 17 \,|\, 9{\cdot}9 - 8{\cdot}8 \,|\, 2{\cdot}32 \div 2{\cdot}9 \,|\, 0{\cdot}001 \times 300 \,|\, 0 \div 9{\cdot}99 \,|\, 1{\cdot}2 \times 10 \times 0{\cdot}2 \,|\, 14{\cdot}3 - 12{\cdot}5 \,\|$

$\|\, 11{\cdot}34 \div 6{\cdot}3 \,|\, \frac{7}{10} \,|\, 0{\cdot}007 \times 200 \,|\, 0{\cdot}24 \div 0{\cdot}12 \,|\, 6{\cdot}69 - 5{\cdot}59 \,|\, 0{\cdot}11 + 0{\cdot}19 \,\|\, 77 \div 770 \,|$

$|\, 0{\cdot}16 \div 0{\cdot}4 \,\|\, 0{\cdot}83 + 0{\cdot}27 \,|\, 2 \times 0{\cdot}7 \,|\, 7{\cdot}15 \div 5{\cdot}5 \,|\, 2{\cdot}9 - 2\frac{3}{10} \,|\, 400 \div 1000 \,|\, 0{\cdot}085 \div 0{\cdot}05 \,\|$

3 Can you find the thickness of a 2p coin in millimetres, correct to one hundredth of a millimetre?

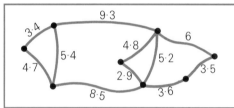

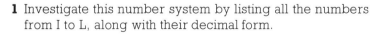

A salesman has to visit all the shops marked by dots. The distances between them are shown to the nearest tenth of a kilometre. Find the shortest route he could take to visit every shop.

The Roman number system
The system is based on these numerals:

Roman	Decimal	Roman	Decimal
I	1	C	100
V	5	D	500
X	10	M	1000
L	50		

A numeral to the *right* is *added* on: II = 2, VI = 6, XXVI = 26.
A numeral to the *left* is *subtracted*: IV = 4, IX = 9, XL = 40.

1 Investigate this number system by listing all the numbers from I to L, along with their decimal form.

2 Try to use Roman numerals to write down the number of days in a year, and also the present calendar year.

3 Can you find some rules for adding numbers written in Roman numerals? Compare them with our decimal number place system and its use of zero.

4 Find out more about Roman numerals and the Roman counting system using reference books and encyclopaedias from your local or school library.

DECIMALS IN ACTION

1A   Decimal notation

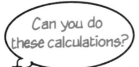

Can you explain the meaning of these numbers as hundreds, tens, etc?

a 34·65
b 875·4
c 0·106

2A   Adding, subtracting, multiplying and dividing decimals

Can you do these calculations?

a 3·68 + 12·5
b 45·8 − 2·9
c 13 − 4·6
d 56·8 × 12
e 9·64 × 1·3
f 101·6 ÷ 8
g 33·48 ÷ 3·6
h 2·485 × 100
i 36·4 ÷ 10

3A, B   Rounding off a decimal

Can you do these?

a Round off to 2 decimal places, then to 1 decimal place:
   (i) 9·083   (ii) 12·685
b Calculate 36·8 ÷ 7, to 2 decimal places.

4B   Changing fractions to decimals

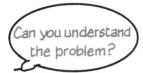

Can you do these?

a Write $\frac{3}{4}$ in decimal form.
b Find $\frac{2}{7}$ as a decimal, rounded off to 3 decimal places.

5A, B   Problems with decimals

Can you understand the problem?

a (i) Calculate the perimeter of this L-shape.

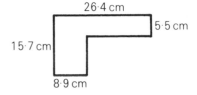

26·4 cm

5·5 cm

15·7 cm

8·9 cm

(ii) Could you have calculated the perimeter if the 5·5 cm and 8·9 cm lengths had not been given?

Can you decide what to do?

b Crumble Wheat breakfast cereal is sold in packs of 8 servings for £1·12 or 12 for £1·56.
How much dearer is each serving of cereal in the smaller pack?

Can you do the calculations?

c Colour films cost £1·95, or £5·74 for a pack of 4. How much do you save by buying a pack of 4?

TILING PATTERNS

On the floor

The tiling pattern on this floor is made from tiles of the **same shape and size**.

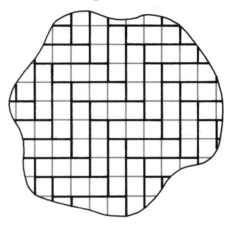

Tiles of the same shape and size are **congruent** to each other.

When fitting tiles together to form a tiling pattern:

Don't leave gaps.
Don't overlap.

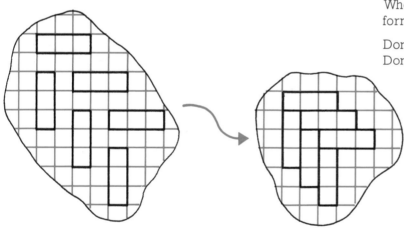

=== *Exercise 1* ===

1A Copy and extend these tiling patterns on squared paper, and then colour them.

One Tile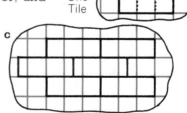

a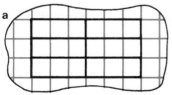

b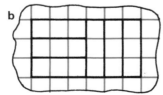

c

2A Using the tile in question **1A**, make up some tiling patterns of your own.

3A These tiling patterns are made from congruent tiles.
Draw one of the tiles from each pattern.

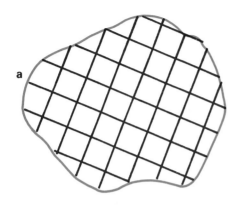

a

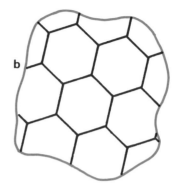

b

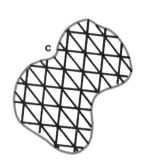

c

Where can you see this tiling?

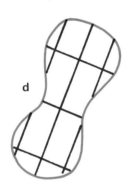

d

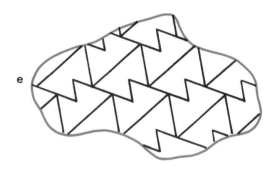

e

4A Copy these patterns on squared paper.
Colour the sets of congruent tiles in each pattern.

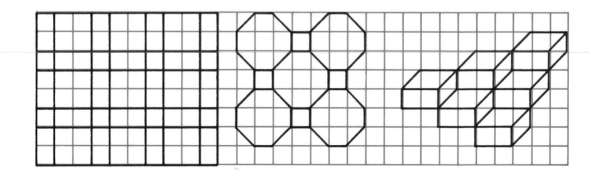

5A Draw one of the tiles from each of these tilings.

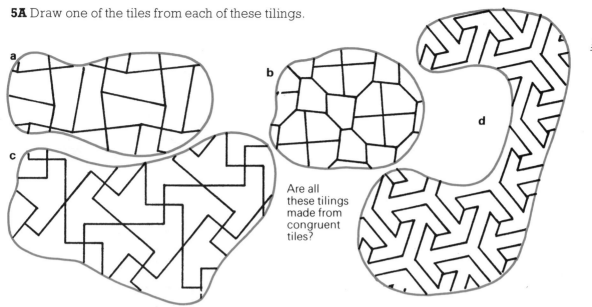

Are all these tilings made from congruent tiles?

6A Here are some floors to tile. Copy them on squared paper, and tile each floor with the tile directly above it.

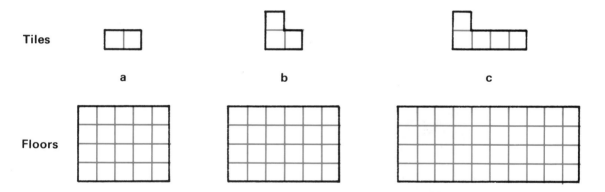

Tiles

a   b   c

Floors

7B Can you tile this floor? Try it.

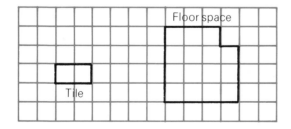

Floor space

Tile

If you cannot, here are some questions to ask yourself.
How many squares make one tile?
How many squares are there in the floor space?
Is the tiling possible? Explain.

8C One domino covers two squares of this chessboard.

    a Can dominoes tile the chessboard?

    b If you take away two of the diagonally opposite corner squares, can the board still be tiled using dominoes?

    c If you take away two of the corner squares on one side of the board can the board be tiled using dominoes?

Copy this tiling pattern on squared paper, and extend it.

You have to colour the tiles so that:

a no two tiles with an edge in common have
* the same colour.

b as few colours as possible are used.

What is the smallest number of colours needed to colour countries in an atlas in this way?

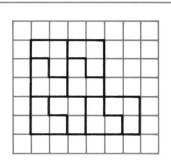

On the desk

Peter Tyler has made four **pentominoes** and has just thought of a fifth. They are all made from five squares joined side to side.

Exercise 2

1A Peter's tiling is not **regular**. It does not make a repeating pattern.

Can you make a repeating tiling pattern with his pentomino? Use a cut-out card pentomino to check your pattern.

2A Can you improve Peter's tiling this time?

3A Try to find one pentomino that Peter did not think of. Can you make a regular tiling pattern with it?

4B

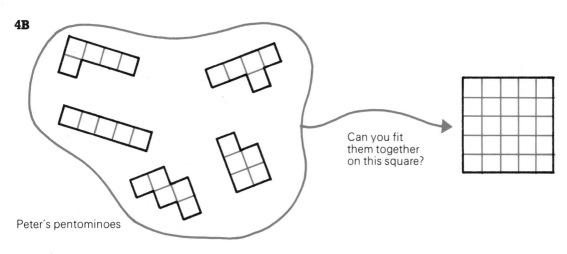

Can you fit them together on this square?

Peter's pentominoes

5C a There are twelve different pentomino shapes. Can you find them all?
  b Try to fit them all on to a 10 by 6 grid of squares.

6C Not all tiles have straight edges. Here are some with curved edges. Can you make tiling patterns with them? Tracing paper might help.

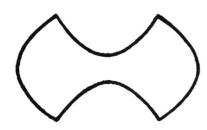

TILING AND SYMMETRY

On the wall

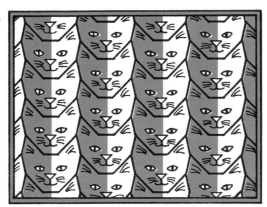 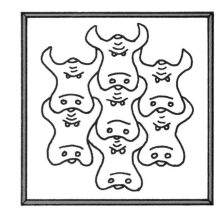

These tiling pictures are made from congruent tiles.
This shows you how to make a plain tile into a picture tile.

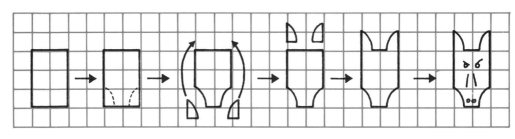

Plain Tiling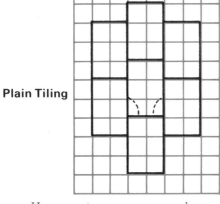

Picture Tiling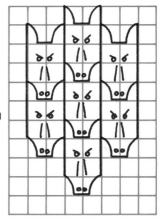

Here are two more examples.

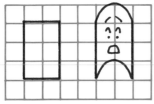

═══════════ *Exercise 3* ═══════════

1A Use the two picture tiles above to draw some picture tilings.

2B Try to make your own picture tilings.

On the grid

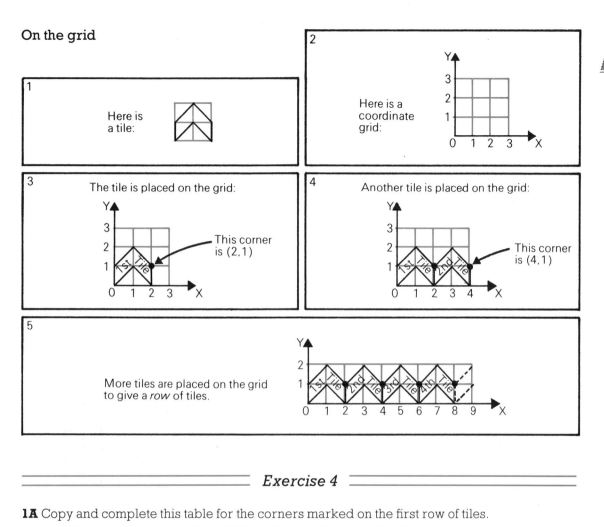

1
Here is
a tile:

2
Here is a
coordinate
grid:

3
The tile is placed on the grid:

This corner
is (2,1)

4
Another tile is placed on the grid:

This corner
is (4,1)

5
More tiles are placed on the grid
to give a *row* of tiles.

Exercise 4

1A Copy and complete this table for the corners marked on the first row of tiles.

1st tile	2nd tile	3rd tile	4th tile	5th tile		50th tile
(2, 1)	(4, 1)	(,)	(,)	(,)	. . .	(,)

2B Can you find

xth tile	?
(,)	

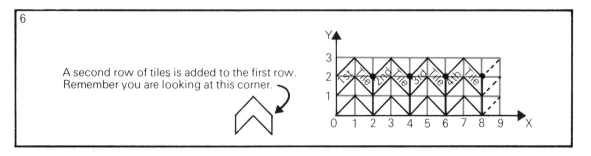

6

A second row of tiles is added to the first row.
Remember you are looking at this corner.

3A Copy and complete this table for the second row of tiles.

1st tile	2nd tile	3rd tile	4th tile	5th tile		50th tile
(2, 2)					...	

4B Can you find

xth tile
(,)

for the second row?

5B Repeat these questions for a third and fourth row.

6C Can you find

xth tile
(,)

for the yth row?

Investigate the coordinates of the top left-hand corner of this tile in the tiling shown alongside.
Try to find the coordinates of the xth tile in the first row, second row, ..., yth row.

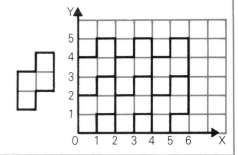

LINE SYMMETRY

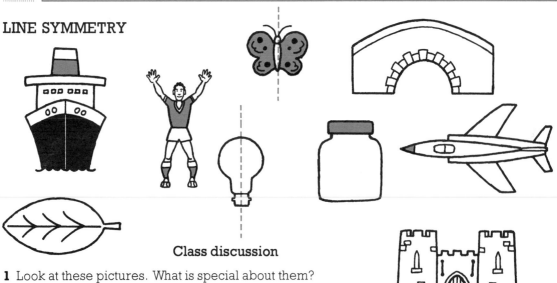

Class discussion

1 Look at these pictures. What is special about them? Why have dotted lines been drawn in two of them? Could this be done for all of them?

2 Trace the butterfly and the light bulb.
Turn the tracings over. Do they fit their outlines?
Try this for two more pictures.
Each picture is 'balanced' about a line, the **line of symmetry** or **axis of symmetry**.

3 Can you see any objects or pictures in the room that have line symmetry? Why do you think so many objects have line symmetry. Sketch some of them.

4 Cut out some pictures from magazines or papers that are symmetrical in this way.

=============== *Exercise 5* ===============

1A Which of these pictures have line symmetry?

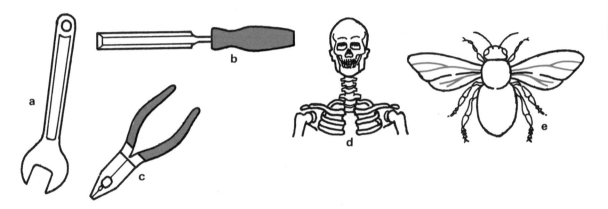

2A Copy these shapes on squared paper. Draw in axes of symmetry, or cut out and fold the shapes to find the axes of symmetry.
How many can you find for each shape?

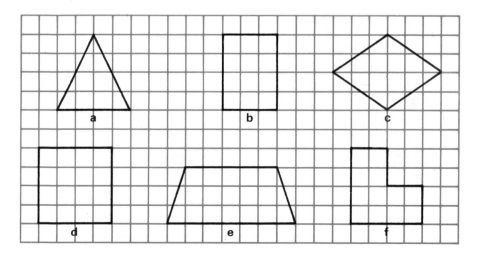

TILING AND SYMMETRY

3A How many lines of symmetry can you find for each of the shapes below?

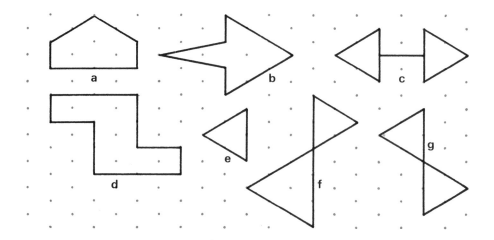

4A Which of these road signs have line symmetry?

T-junction

Low flying aircraft

Cattle

End of
speed limit

Keep right

Turn left

5A Think of all the letters of the alphabet.
Make up lists of the capital letters that
should arrive at each barrel. Draw some
of the letters, showing lines of symmetry.

6A Can you think of everyday objects that have more than one axis of symmetry?

146

7B How many lines of symmetry do these have? (Use tracing paper if you need to.)

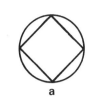
a

b

c

d

e

f

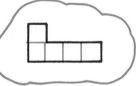
g

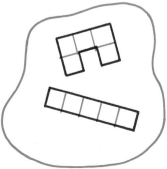
h

i

8B Show how to fit a pair of each of these pentominoes together to make a shape with an axis of symmetry.

a

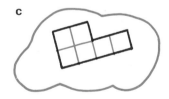
b
c

9B Fit each of these pairs of different pentominoes together to make a shape with an axis of symmetry. Show each one in a diagram.

a

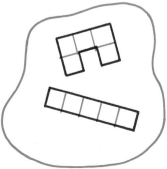
b

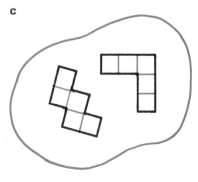
c

TILING AND SYMMETRY

147

1 Many leaves are symmetrical about a line. Mount a collection of these.
2 Many sports, including football, cricket, tennis and gliding, make use of symmetry in the design of their equipment and pitches. Prepare a report on one of these.

On reflection

You can make symmetrical patterns by reflection in a mirror, like this:

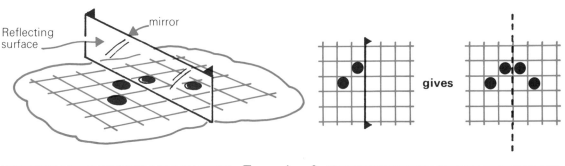

gives

=== *Exercise 6* ===

1A Copy these diagrams on squared paper, showing the dots and their reflections. A mirror can help.

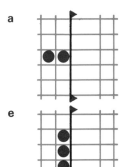

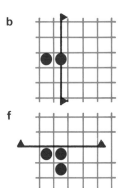

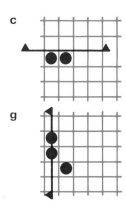

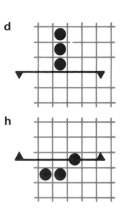

2B Here are some harder diagrams.

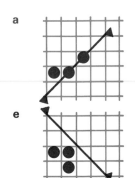

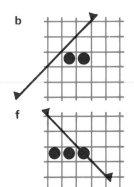

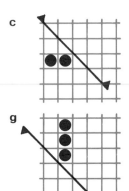

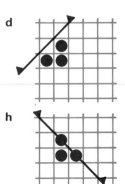

Cut it out

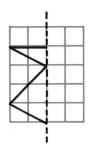

Starting shape   **Completed symmetrical shape**

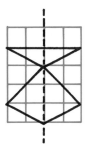

gives

The result can be checked by drawing and cutting out.

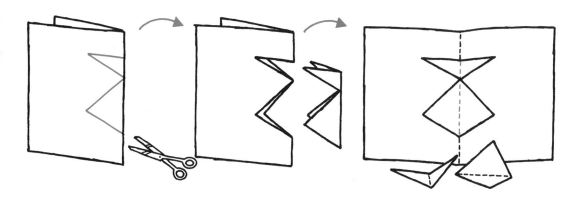

================ *Exercise 7* ================

1A Copy these shapes on squared paper. Then complete them so that the dotted lines are axes of symmetry. Check your result by cutting the shape out and folding it along the axis of symmetry.

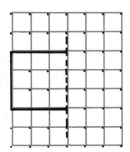

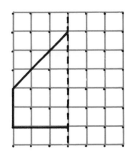

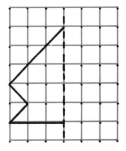

2A Repeat question **1A**, starting with some shapes of your own. Include everyday shapes like houses, faces, tennis rackets, aircraft etc.

TILING AND SYMMETRY

Starting shape

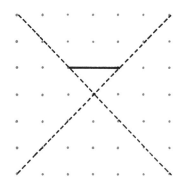

Completed symmetrical shape

gives

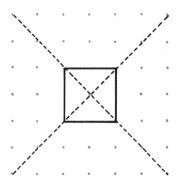

To check this result, fold and cut like this:

Second fold

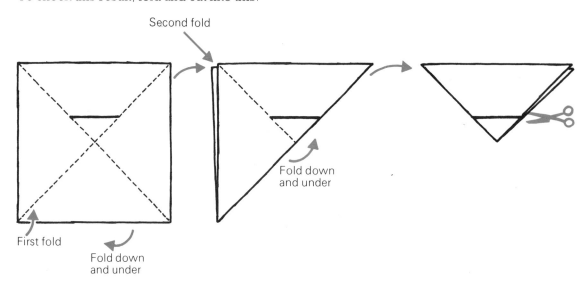

Fold down
and under

First fold

Fold down
and under

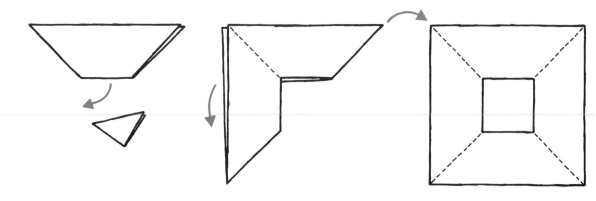

Unfold second fold   Unfold first fold

3B Copy these shapes on squared paper. Complete them so that both the dotted lines are axes of symmetry. Check your result by cutting the shape out of folded paper.

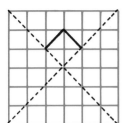 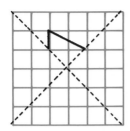 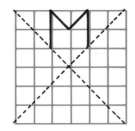 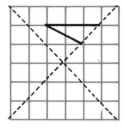

4C Copy and complete these diagrams so that all the dotted lines are axes of symmetry. How can you fold the paper so that your results can be checked?

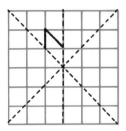 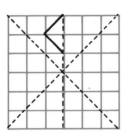 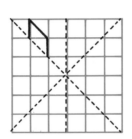 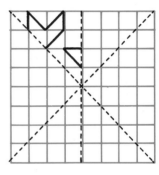

What's the point?

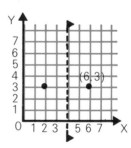

The *image* of the point $(2, 3)$ in the mirror is the point $(6, 3)$. The *y*-coordinate, 3, stays the same.

=========== *Exercise 8* ===========

1A Copy the first diagram above onto squared paper. Plot the points $(1, 4)$, $(3, 6)$, $(0, 2)$, $(3, 0)$, $(2, 5)$. Now plot their images.
   Which coordinate changes? Which coordinate stays the same?

In the following questions, plot the points on squared paper. Reflect them in the mirrors (the dotted lines). Then complete the tables.

2A

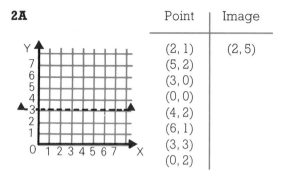

Point	Image
(2, 1)	(2, 5)
(5, 2)	
(3, 0)	
(0, 0)	
(4, 2)	
(6, 1)	
(3, 3)	
(0, 2)	

Which coordinate stays the same?

3A

Point	Image
(2, 1)	
(4, 2)	
(5, 4)	
(0, 3)	
(3, 3)	
(6, 5)	
(1, 3)	

Two-sided mirror

Which coordinate stays the same?

4A

Point	Image
(1, 1)	
(3, 0)	
(5, 4)	
(4, 3)	
(0, 3)	
(2, 2)	
(6, 2)	

5A

Point	Image
(2, 1)	
(3, 1)	
(2, 4)	
(4, 4)	
(5, 1)	
(2, 5)	
(0, 4)	
(3, 0)	
(6, 6)	

Look at each point and its image. What do you notice when you compare them? Copy and complete: The image of (a, b) is $(...., ...)$.

6B Copy and complete this table.
Compare the y-coordinate of each point with the x-coordinate of its image. Write a sentence about what you notice. What is the image of the point (a, b)?

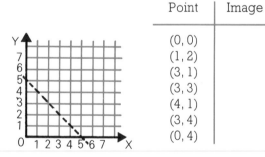

Point	Image
(0, 0)	
(1, 2)	
(3, 1)	
(3, 3)	
(4, 1)	
(3, 4)	
(0, 4)	

7C In this question choose your own points. Find their images, and try to find patterns. Investigate the image of (a, b) in each case.

a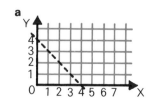

b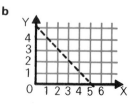

TURN SYMMETRY

Upside down

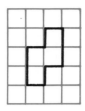

TURN YOUR PAGE UPSIDE DOWN.

Does this shape look the same?
Turn the page back again to check.

This shape has half-turn symmetry.

TURN YOUR PAGE SIDEWAYS.

Does this shape look the same? Turn the page back again to check.

This shape has quarter-turn symmetry.

=== *Exercise 9* ===

1A Find out whether each shape below has quarter-turn or only half-turn symmetry.

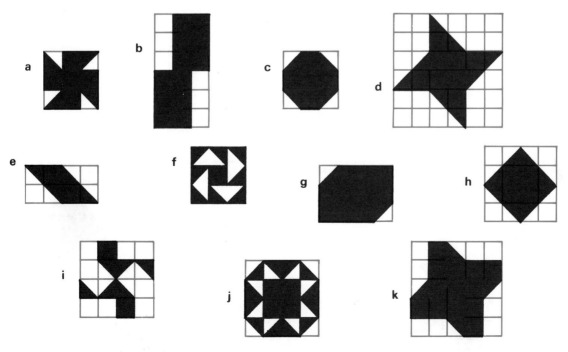

2A Which of these symbols have:

    a line symmetry   **b** half-turn symmetry   **c** quarter-turn symmetry?

3A Make up lists of the capital letters of the alphabet that arrive at each of these barrels.

4B What letters arrive at these barrels?

5B This pentomino has quarter-turn symmetry.
Are there other pentominoes with quarter-turn symmetry?
Find some pentominoes with half-turn symmetry.

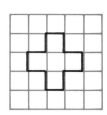

6C Investigate pairs of pentominoes which fit together to make shapes with half-turn symmetry.

Not without trace

═══════════════════ *Exercise 10* ═══════════════════

1A Trace each of these shapes. Put the point of your pencil on the dot under the tracing. Turn the tracing round to check that the shape has half-turn symmetry. **The dot is the centre of symmetry.**

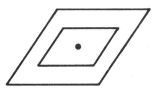  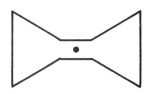

2A Copy the drawings below on squared paper.
Complete the drawings so that the dot becomes the centre of half-turn symmetry.

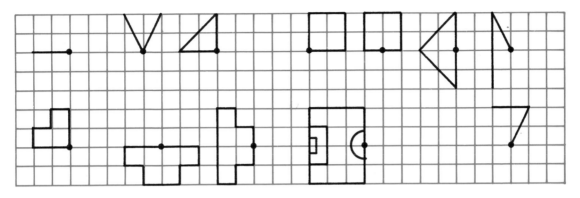

3B Which of the following statements are true and which are false?

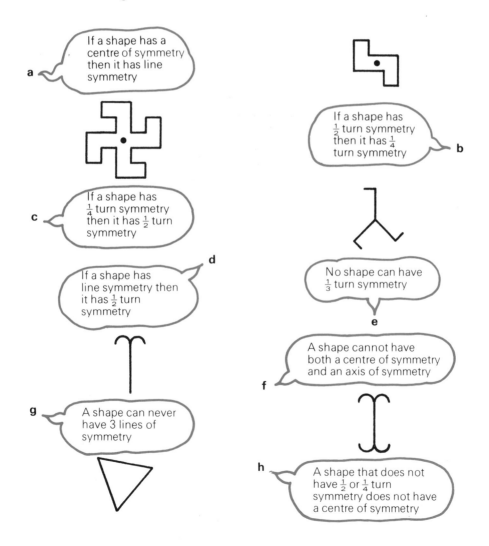

a If a shape has a centre of symmetry then it has line symmetry

b If a shape has $\frac{1}{2}$ turn symmetry then it has $\frac{1}{4}$ turn symmetry

c If a shape has $\frac{1}{4}$ turn symmetry then it has $\frac{1}{2}$ turn symmetry

d If a shape has line symmetry then it has $\frac{1}{2}$ turn symmetry

e No shape can have $\frac{1}{3}$ turn symmetry

f A shape cannot have both a centre of symmetry and an axis of symmetry

g A shape can never have 3 lines of symmetry

h A shape that does not have $\frac{1}{2}$ or $\frac{1}{4}$ turn symmetry does not have a centre of symmetry

1 Draw five 4 × 4 squares.
Colour them so that the patterns have 1, 2 or 4 axes of symmetry, or quarter or half-turn symmetry (one drawing for each).

2 Draw a circle with radius 6 cm long.
Mark twelve equally spaced points on the circle.
Join one of the points to all the other points.
Join the next point to all the other points.
Repeat this until all the points have been joined.
How many different kinds of symmetry can you find in this 'magic rose' picture?

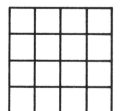

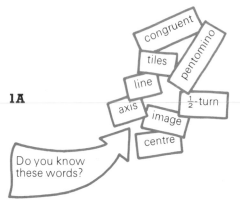

CHECK-UP ON **TILING AND SYMMETRY**

1A

Do you know these words?

congruent tiles pentomino line axis ½-turn image centre

A · B

a *Copy and complete:*

These two shapes are _____.
This _____ has _____ symmetry.
The dotted line is the_____ of symmetry.

It does not have _____ symmetry and so has no _____ of symmetry.

b A is the _____ of B in the dotted line.

2A

Can you extend a tiling?

a Copy and complete this tiling.

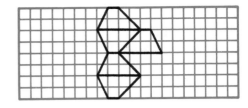

Can you make up a regular tiling pattern?

b Make up a regular tiling pattern with this tile.

Can you find symmetry in shapes?

c These are Egyptian symbols. Copy the ones that have symmetry, and say what kind of symmetry they have.

3B

Can you complete symmetrical shapes?

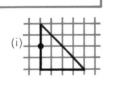

a Complete these two drawings.
In (i) make the dot the centre of symmetry.
In (ii) make the dotted lines axes of symmetry.

Can you find the image of a point?

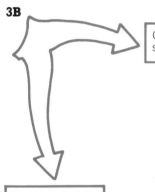

b Find the images of (2, 1) and (5, 4) in the dotted line.
Find the image of $(x, 3)$ in the same line.

4C

Investigate:

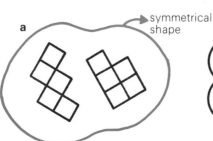

a symmetrical shape **b**

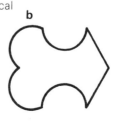

10 MEASURING TIME

The calendar

The moon takes nearly one month to move round the earth.

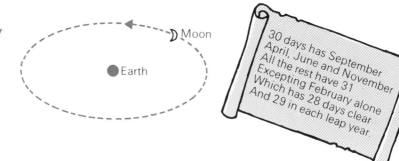

Moon

Earth

30 days has September
April, June and November
All the rest have 31
Excepting February alone
Which has 28 days clear
And 29 in each leap year.

TIME AND TEMPERATURE

===== *Exercise 1* =====

1A List the months of the year under the headings 28 days, 29 days, 30 days, 31 days.

2A How many days are there altogether in the last three months of the year?

3A a Copy and complete this page from a calendar.
  b What day of the week was 17th June?
  c What was the date of the third Tuesday in June?

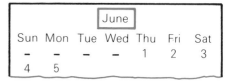

			June			
Sun	Mon	Tue	Wed	Thu	Fri	Sat
–	–	–	–	1	2	3
4	5					

4A a A holiday starts on a Monday and lasts for 9 days. On what day does it finish?
  b From the first day of term to the last is 80 days including Saturdays and Sundays. Term begins on a Monday. On what day does it end?

5A Spring starts on March 20th and ends on June 20th. To calculate the length of Spring, copy and complete:

$$\begin{array}{ll} \text{March: 20 to 31} = 12 \text{ days} \\ \text{April:} \\ \text{May:} \\ \text{June:} \qquad\qquad 20 \\ \hline \text{TOTAL} \qquad \text{days} \\ \hline \end{array}$$

6A Calculate the number of days (including both dates, as in question **5A**) from:

  a January 1st to February 24th

  b June 28th to August 18th.

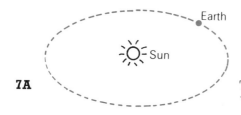

Earth

Sun

7A

The earth moves round the sun in about $365\frac{1}{4}$ days. What happens to the $\frac{1}{4}$ day?

8A Use the flowchart to find out which
of these years are leap years:

    a 1990   **b** 1992   **c** 1998
    d 2000   **e** 2100   **f** 1789

9A a Is the present year a leap year?
    b If not, when was the last leap year?

10A

How many runs has he scored?

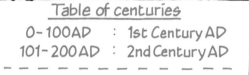

Table of centuries

0 – 100 AD	: 1st Century AD
101 – 200 AD	: 2nd Century AD
1901 – 2000 AD	: 20th Century AD
2001 – 2100 AD	: 21st Century AD

To find a leap year

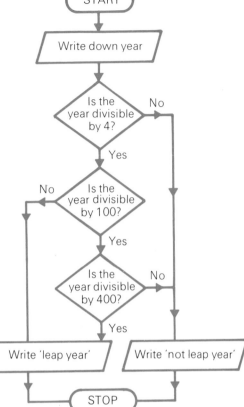

11A Make a list of these dates and their centuries:

Columbus discovered America 1492 AD

Romans left Britain 410 AD

First man landed on the moon 1969 AD

Battle of Bannockburn 1314 AD

St. George, Patron Saint of England died 303 AD

Great Fire of London 1666 AD

Battle of Hastings 1066 AD

Shakespeare born 1564 AD

12A How many centuries are there in the period:
    a 1850 AD to 2050 AD   **b** 991 AD to 1991 AD?

13A Write down the numbers that go in these spaces:

. . . seconds in a minute	. . . or . . . or . . . or . . . days in a month
. . . minutes in an hour	. . . or . . . days in a year
. . . hours in a day	. . . weeks in a year
. . . days in a week	. . . months in a year
. . . days in a fortnight	. . . years in a decade
	. . . years in a century

14B With the help of the flowchart in question **8A**, write one or two sentences to explain how to find out whether or not a year is a leap year.

TIME AND TEMPERATURE

Clock time
'It's 4 o'clock.'
But is it 4 am ('ante meridiem' means before noon),
or 4 pm ('post meridiem' means after noon)?
We can't tell from the 12 hour clock.

═══════════════ *Exercise 2* ═══════════════

1A What are the two possible times shown by each of these clocks?

 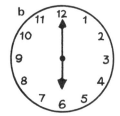  

2A Write these afternoon times in figures:
  a half past one   **b** twenty past two   **c** quarter to three   **d** twenty-five to five.

3A How long is it from:
  a 10 am until noon   **b** noon until 5 pm   **c** 9 am until 3 pm   **d** 2 am until 11 am
  e 5 am until 3 pm   **f** 7 am until midnight   **g** 10 pm until 8 am   **h** 3.30 pm until 11.30 pm?

4A Bill waited for a bus from quarter to five until ten past five. How long was this?

5A Grace has to be at the airport at least an hour and a half before her plane is due to take off at 10.15 pm. When should she be there?

6A Alex hands in his shoes at 1.20 pm.
  When will they be ready?

7A Lois is going to roast a chicken. She has to allow 20 minutes for each 500 g, plus 15 minutes over.
  a How long will it take to roast a chicken weighing 2 kg?
  b If the chicken is put in the oven at 3 pm, when will it be ready?
  b If the chicken has to be ready at 6 pm, when should Lois put it in the oven?

8A a On August 1st the sun rose at 5.22 am and set at 9.24 pm. How long was it above the horizon? Copy and complete:

  5.22 am – midday =   6 hours 38 minutes
  midday – 9.24 pm = . . . hours 24 minutes
  ─────────────────────
       . . . hours . . . minutes (38 + 24 = 62 minutes = 1 hour 2 minutes)
  ─────────────────────

  b On the same day the moon rose at 10.21 pm and set at 5.16 am. For how long could the moon be seen?

9A a A bus service between two towns runs every 40 minutes. The first bus leaves the bus station at 7.30 am. When will the next two buses leave?

b The journey takes 1 hour 25 minutes. Write down the arrival times of these three buses.

c The third bus reaches its destination 25 minutes late. What time does it arrive?

10B Here are some world record times for running the mile:

4 minutes	6·4	seconds	by Sydney Wooderson	in	1937
3 ,,	59·4	,,	,, Roger Bannister	,,	1954
3 ,,	48·40	,,	,, Steve Ovett	,,	1981
3 ,,	47·33	,,	,, Sebastian Coe	,,	1981
3 ,,	46·31	,,	,, Steve Cram	,,	1985

What was the difference in time between:

a Cram and Coe **b** Coe and Ovett

c Bannister and Wooderson **d** Cram and Wooderson?

The 24 hour clock

Many bus, train and plane timetables use the 24 hour clock, so that the same hour won't be used twice a day. This clock runs from midnight to midnight.

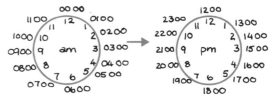

03 00 hours = 3 am 14 00 hours = 2 pm
06 30 hours = 6.30 am 19 30 hours = 7.30 pm
10 00 hours = 10 am 23 00 hours = 11 pm
(Subtract 12 for pm time).

Midnight *Midday* *Midnight*

←———am———→ ←———pm———→

00 00 hours ———→ 12 00 hours ———→ 24 00 hours

=== *Exercise 3* ===

1A Write down the times on these clocks: **a** in 24 hour time **b** in 12 hour (am/pm) time.

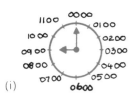

(i)

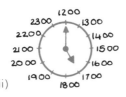

(ii)

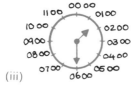

(iii)

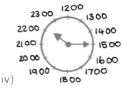

(iv)

2A Copy and complete this table:

12 hour clock	4 am	10 am	2 pm	5 pm				
24 hour clock					03 00 hours	16 00 hours	11 00 hours	21 00 hours

3A What times, am or pm, are these watches showing?

a    b    c    d

4A A bus leaves the bus station at 18 25 hours, and reaches its destination at 20 10 hours. How long did the journey take? Copy and complete this calculation:

$$18\,25\text{ hours to } 19\,00\text{ hours} = \qquad 35\text{ minutes}$$
$$19\,00\text{ hours to } 20\,10\text{ hours} = \ldots\text{ hour} \ldots\text{ minutes}$$
$$\text{Total time} = \underline{\qquad\qquad\qquad}$$

5A A train leaves at 14 35 hours and arrives at 16 20 hours. How long does the journey take?

6B Ian is on holiday at the seaside. High tide today is at 15 30 hours. The next high tide will be in 12 hours and 24 minutes. What time will that be?

7B How long is it between these times?

    a 15 40 hours and 17 00 hours   **b** 08 15 hours and 10 50 hours

    c 21 46 hours and 23 35 hours   **d** 22 05 hours and 03 20 hours next day

8B Kate wants to set the timer on her video recorder to record a programme starting at 7.35 pm and lasting for 2 hours 45 minutes. If the recorder uses a 24 hour clock what times should be set?

1 Look at your watch. Look at it again after you think 30 seconds have passed. If your guess wasn't accurate, try counting 'thousand-one', 'thousand-two', . . . for the seconds.

2 How long does it take you to come to school? Keep a record for a week, and calculate the average time by dividing the total time by 5.

3 Calculate the number of minutes since you were born.

4 If March 3rd is a Monday one year, what day could it be the following year?

5 Find out about British Summer Time and its relation to Greenwich Mean Time.

6 Find out how the months got their names and their number of days.

7 Investigate time zones which define the time in nearly every country in the world.

Andrew's timetable

	8.55 –9.10	Period 1 9.10–10.10	Period 2 10.10–11.10		Period 3 11.25–12.30		Period 4 1.30–2.35	Period 5 2.35–3.40
Monday	A	Science	French	B	Geography	L	English	Maths
Tuesday	S S	History	Music	R	Home Economics	U	French	English
Wednesday	E M	Maths	Art	E	Science	N	Geography	P.E.
Thursday	B L	English	Technical Education	A	French	C	Maths	Science
Friday	Y	P.E.	English	K	Maths	H	Art	History

========== *Exercise 4* ==========

1A Period 1 lasts 1 hour, from 9.10 am until 10.10 am. How long do periods 2, 3, 4 and 5 last?

2A How long does the Assembly take?

3A Write down all the subjects Andrew studies. How many are there?

4A a How many periods of English does he have in a week?
  b Count the number of hours and minutes of English in a week.

5A Repeat question **4A** for P.E. and for Science.

6B Calculate the number of hours and minutes there are in:
  a a school day   **b** a school week. Include the time for assembly, but not for break or lunch.

7B a Answer questions **3A–6B** for your own timetable.
  b What subjects would you like to have: (i) more of (ii) less of?
    Give reasons for your answer.

8B Design a class timetable, using 8 periods of 35 minutes or 7 periods of 40 minutes daily, or another suitable pattern.

Television programmes

First thing on Saturday morning Donna looked at the day's TV programmes.

Todays Viewing

BBC1

7.10 OPEN UNIVERSITY.
8.30 THE SATURDAY PICTURE SHOW.
10.40 TROOPING THE COLOUR.
12.15 GRANDSTAND introduced by Bob Wilson, including 1.0 News Summary; Weather; Cricket; Tennis; Final Score.
5.10 NEWS; WEATHER.
5.20 SPORT.
5.25 THE NEW ADVENTURES OF WONDER WOMAN.
6.15 TERRY AND JUNE.
6.45 FILM: "Catch Me A Spy."
8.10 DYNASTY: "That Holiday Spirit."
9.00 THE VAL DOONICAN MUSIC SHOW.
9.45 NEWS; WEATHER.
10.00 THE ROYAL INTERNATIONAL HORSE SHOW.
11.10 FILM: "The War Between Men and Women."
12.55 WEATHER; CLOSEDOWN.

BBC2

6.25 OPEN UNIVERSITY.
3.10 FILM: "The Savage Guns."
4.30 CRICKET: FIRST TEST.
6.20 DANCE INTERNATIONAL.
6.30 TROOPING THE COLOUR. Highlights of this morning's military parade.
7.55 NEWS AND SPORT; WEATHER.
8.10 SATURDAY REVIEW.
9.00 FILM: "The Godfather Part II."
12.15 CRICKET: FIRST TEST. Richie Benaud introduces highlights of England v Australia.
12.45 INTERNATIONAL TENNIS. Highlights of today's semi-final matches.
1.30 CLOSEDOWN.

CHANNEL 4

1.05 CHIPS' COMIC.
1.30 LISTENING EYE.
2.00 FILM: "Wilson."
4.50 THE WEEKEND STARTS HERE.
5.05 BROOKSIDE.
6.00 THE MAX HEADROOM SHOW.
6.30 BABBLE.
7.00 NEWS SUMMARY and WEATHER followed by 7 DAYS.
7.30 UNION WORLD.
8.00 TALES FROM A LONG ROOM.
8.15 SANNE. The second episode of the Dutch drama serial.
8.45 MY WORLD AND WELCOME TO IT.
9.15 TO BE A YELLOWBELLY . . .?
10.00 OCTOPUS – POWER OF THE MAFIA.
11.00 GOLF – THE US OPEN CHAMPIONSHIP.
1.00 THE PAUL HOGAN SHOW.
1.30 CLOSEDOWN.

ITV

9.25 CARTOON TIME.
9.35 SCOOBY DOO.
10.00 No. 73.
11.20 CHIPS.
12.15 WORLD OF SPORT: Wrestling; ITN News; Basketball; Rallying; Golf; The ITV Six; Speedway; Athletics; Results.
5.00 NEWS.
5.05 CONNECTIONS.
5.35 FILM: "The Navy Lark."
7.00 THE COMEDIANS.
7.30 THE PRICE IS RIGHT.
8.30 HUNTER.
9.30 NEWS.
9.45 TALES OF THE UNEXPECTED.
10.15 FILM: "Attica – Story of a Prison Riot."
12.00 LATE CALL. Rev. John Bell of the Iona Community.
12.05 CLOSEDOWN.

Exercise 5

1A How many hours of viewing are there from start to closedown on: **a** BBC1 **b** ITV?

2A Donna is very keen on sport. How many hours of sport could she watch on: **a** BBC1 **b** ITV? List the programmes, and the time they last, in a table.

3A She also likes films. How much time does each channel give to films?

4A Her father always watches the News. How much time is given to news on:

a BBC1 **b** ITV **c** Channel 4?

5A Donna wants to record Dynasty, The Royal International Horse Show and International Tennis. Can she do this on a 3-hour video tape?

6A You decide to spend the whole evening watching television. Make a list of the programmes and channels you would choose from tea-time till bedtime.

Programme planner

=== *Exercise 6* ===

1B Plan a day's programmes for Channel 100! You have a free choice, except that News must be shown at 1 pm, 6 pm and 10 pm, and 'Wakey, Wakey, Britain' must begin at 7 am. Channel 100 closes down at midnight.
Here are the programmes and the times they last:

News at One	—30 minutes	News at Six	—30 minutes
News at Ten	—30 minutes	Wakey, Wakey, Britain	—3 hours
Stars and Planets	—1 hour	Cartoons	—20 minutes
Film	—2 hours	Film	—1 hour 30 minutes
World Religions	—40 minutes	Football Highlights	—20 minutes
Chat Show	—1 hour	The Comedians	—1 hour
For Schools	—1 hour	Criss-Cross Quiz	—30 minutes
Murder and Mystery	—1 hour	Politics To-Day	—40 minutes
Pop Music	—1 hour	Our Zoos	—30 minutes

If necessary you can 'fill in' with travel films, each lasting 10 minutes.

A train timetable

Some people are interested in train timetables. Fiona got this one in her local station. It shows the times of trains from Edinburgh to Stirling and Perth on Sundays.

Edinburgh	09 15	10 30	12 30	14 30	16 20	16 30	17 15	17 55	18 45	19 45	20 45	22 15	23 25
Haymarket	09 18	.....	.....	.....	16 23	.....	17 18	17 58	18 48	19 48	20 48	22 18	23 28
Linlithgow.......	.....	.....	.....	.....	.....	.....	17 36	.....	19 06	20 06	21 06	22 36	.....
Polmont	.....	.....	.....	.....	.....	.....	17 43	.....	19 13	20 13	21 13	22 43	.....
Falkirk	09 41	11 03	13 03	15 03	.....	17 03	17 49	18 20	19 19	20 19	21 19	22 49	23 53
Larbert.........	09 58	11 08	13 08	15 08	.....	17 08	17 55	.....	19 25	20 25	21 25	22 55	00 01
Stirling	10 08	11 19	13 19	15 19	.....	17 19	18 05	18 33	19 35	20 35	21 35	23 05	00 11
Bridge of Allan...	.....	11 24	13 24	15 24	.....	17 24	18 10	.....	19 40	20 40	21 40	23 10	.....
Dunblane........	.....	11 31	13 31	15 31	.....	17 31	18 18	18 41	19 48	20 48	21 48	23 18	00 21
Gleneagles......	.....	.....	.....	.....	.....	.....	.....	18 53	.....	.....	.....	.....	.....
Perth..........	10 48	.....	.....	.....	17 50	.....	.....	19 09	.....	.....	.....	.....	00 54

=== *Exercise 7* ===

1A a Which stations does this train stop at?
b How long does it take to travel to Perth?

The train now standing at platform 7 is the 09 15 for Perth

2A When does the first train for Polmont leave Edinburgh?

3A When does the last train for Bridge of Allan leave Falkirk?

4A If you had to be in Stirling by 8 pm, what is the latest train that you could take from Edinburgh?

TIME AND TEMPERATURE

5A How many trains travel on Sundays from:
  a Edinburgh to Perth **b** Stirling to Dunblane?

6A How many trains leave Edinburgh for Stirling or Perth:
  a in the morning **b** between noon and 6 pm **c** after 6 pm?

7A a At which station on the line does only one train from Edinburgh call on a Sunday?
  b At which towns do most trains stop on a Sunday?

8B How long does it take from Edinburgh to Perth by:
  a the fastest train **b** the slowest train?

9C The distance from Edinburgh to Perth is 42 miles. Calculate the average speed of the train which leaves Edinburgh at 16 20 hours; that is, the distance travelled divided by the time taken.

Flight times

============================= *Exercise 8* =============================

1A Travel British Airways Worldwide! London to Oslo daily flights.

London Airport	Depart	Arrive	Aircraft	Flight Number
Heathrow	08 25	10 20	Boeing 757	BA642
Gatwick	10 35	13 45	Douglas DC9	SK516
Heathrow	10 45	12 45	Douglas DC9	SK514
Heathrow	14 05	16 00	Boeing 757	BA644
Heathrow	17 50	19 50	Douglas DC9	SK512

Departure and arrival times are both in British Summer Time (BST)

  a How long does the quickest flight take?
  b How long does the slowest flight take?
  c If you have to be in Oslo by 5 pm (BST), which is the latest flight you can take from London?

2B London Heathrow to New York—daily flights in summer.

Depart	Arrive	Aircraft	Flight No.
10 30	09 20	Concorde	BA193
11 00	13 30	Boeing 747	BA175
13 15	15 45	Boeing 747	BA177
18 00	16 50	Concorde	BA195
18 30	21 00	Boeing 747	BA179

Departure—British time
Arrival—New York time
(5 hours behind British time).

  a When it is 10 30 hours in London, what is the *local* time in New York?

  See the notice on the right.

  b How long does Concorde take to fly to New York?
  c How much longer does a Boeing 747 take?

3C Calculate the average speeds of the Boeing and Concorde on the flights from London to New York. The distance is 3500 miles. Does either average speed exceed the speed of sound (760 miles per hour)?

MEASURING TEMPERATURE

Feeling the heat
Temperature is measured in degrees Celsius (°C), using a thermometer.
How many degrees are there between the temperatures at which water freezes and boils?
What is the temperature in your classroom?

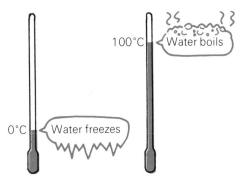

100°C ⟨Water boils⟩

0°C ⟨Water freezes⟩

Here are the temperatures round the world one day in July, in degrees Celsius.

Alicante	31°C	London	20°C	Oslo	18°C
Athens	36°C	Los Angeles	30°C	Palma	30°C
Blackpool	19°C	Miami	30°C	Paris	24°C
Corfu	33°C	Moscow	25°C	Rome	33°C
Edinburgh	19°C	Naples	32°C	Sydney	17°C
Hong Kong	31°C	New Delhi	34°C	Toronto	24°C
Las Palmas	25°C	New York	32°C	Venice	26°C

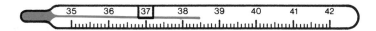

Exercise 9

1A a Which place was warmest?
b Which place was coolest?
c Calculate the difference in temperature between these two places.

2A Which places had temperatures of 30° or more?

3A What was the difference in temperature between:
a Blackpool and Hong Kong   **b** New York and Moscow   **c** Miami and Palma?

4A Peter was not feeling well. His mother took his temperature with this thermometer.

35   36   37   38   39   40   41   42

a What was his temperature?
b Why is there a box round the 37 on the scale?
c Why does the scale only go from 35°C to 42°C?

5A What are the temperatures on these thermometers?

35   36   37   38   39   40   41   42   35   36   37   38   39   40   41   42

a   b

TIME AND TEMPERATURE

6A Different scales are used on different kinds of thermometer.
Look at these very carefully. Then write down the temperatures marked **a, b, c, d, e, f, g**.

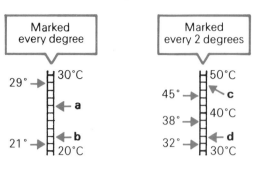
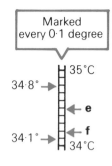
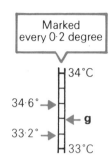

7A What are the temperatures on these thermometers (all in °C)?

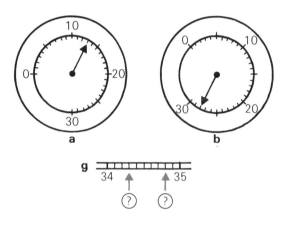
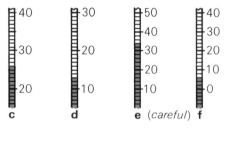
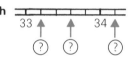

8A The oven tap on a gas cooker is marked
'OFF, $\frac{1}{4}$, $\frac{1}{2}$, 1, 2, 3, . . . , 9'. These settings
and their temperatures are shown in the
table.

  a A recipe for a cake says that the oven
  temperature should be 190°C. What
  setting is this?

  b Meringues are baked at a setting of $\frac{1}{2}$.
  What temperature is this?

  c Estimate the temperatures of these
  settings.

Setting	Temperature °C
$\frac{1}{4}$	110
$\frac{1}{2}$	130
1	140
2	150
3	170
4	180
5	190
6	200
7	220
8	230
9	240

9B Copy and complete this table.

First temperature	Second temperature	Rise in temperature	Fall in temperature
64°C	75°C		—
52°C	43°C	—	
32°C		9°	—
	25°C	6°	—
28°C		—	7°
	13°C	—	8°
12·5°C	15·8°C		
21·4°C	16·7°C		
13·6°C		3·5°	—
	21·3°C	—	4·6°

10C Some thermometers used in greenhouses or in freezers show the maximum and minimum temperatures since they were last set.

For each of the following write down (in °C):

a the present temperature   **b** the maximum temperature   **c** the minimum temperature

d the difference in temperature between the maximum and minimum readings.

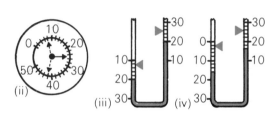

Can you say where else thermometers like these might be useful?

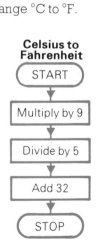

Temperatures used to be measured in degrees Fahrenheit. This scale is still used sometimes.

The weatherman reads the temperature. Here is his flowchart to change °C to °F.

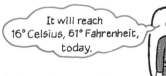

It will reach 16° Celsius, 61° Fahrenheit, today.

Celsius to Fahrenheit

START

Multiply by 9

Divide by 5

Add 32

STOP

1 Use the flowchart to change these temperatures to °F:

   a 20°C   **b** 40°C   **c** 0°C   **d** 100°C

2 Investigate the temperatures on both Celsius and Fahrenheit scales for:

   a a warm summer day   **b** a cold winter day

   c the lounge in your home   **d** a hot bath

   e your body   **f** the classroom

TIME AND TEMPERATURE

1A The calendar

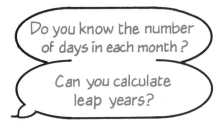

a How many days are there in July?
b Which of these are leap years?
   (i) 1984   (ii) 1362   (iii) 1500
c How many days are there between 24th September and 5th November, counting both dates?

2A, B The 12 and 24-hour clocks

a pm lasts from .... until ....
b How long is it from 9.28 pm until 2.14 am the next morning?
c What name do we give to:
   (i) 12 00 hours   (ii) 24 00 hours?
d What is 4.15 pm in 24-hour time?
e How long is it from 13 48 until 21 05 on the same day?

3A Timetables

| **Depart** | 09 30 | 13 30 | 21 30 |
| **Arrive** | 11 45 | 15 50 | 22 30 |

How long do the longest and shortest journeys take?

4A, B Temperature

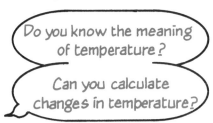

a Arrange these temperatures in order, from coolest to warmest:
   55°C, 100°C, 0°C, 19°C, 21°C.
b Would you enjoy sunbathing in a temperature of: (i) 45°C   (ii) 20°C?
c What is the rise in temperature from 12·7°C to 18·3°C?

What temperatures are shown below?

5A, B Reading thermometers

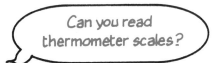

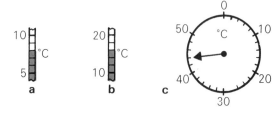

REMINDERS

Reminder 1.   Horses and carts

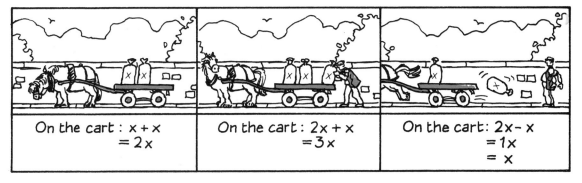

On the cart : x + x
        = 2x

On the cart : 2x + x
        = 3x

On the cart: 2x − x
        = 1x
        = x

═══════════════════ *Exercise 1* ═══════════════════

Write in a shorter form:

1A $a+a$ **2A** $b+b+b$ **3A** $c+c+c+c$ **4A** $2n+n$

5A $3m+m$ **6A** $2x+2x$ **7A** $a+3a$ **8A** $3x+2x$

9A $2x-x$ **10A** $2y-y$ **11A** $3a-a$ **12A** $4x-x$

13A $3t-2t$ **14A** $4u-3u$ **15A** $4x-2x$ **16A** $3w-w$

17B $3x+2x-x$ **18B** $y+2y-y$ **19B** $2k+4k-k$ **20B** $6n-4n+3n$

Reminder 2.   A cover up

In Chapter **6** you solved equations by the 'cover up' method.

For example, to solve $2x+3 = 9$

$$2x = 6$$
$$x = 3$$

$x = 3$ is the solution of the equation.

$\boxed{6} + 3 = 9$

═══════════════════ *Exercise 2* ═══════════════════

Use the 'cover up' method to solve these equations:

1A $x+5 = 7$ **2A** $y-3 = 6$ **3A** $n+4 = 4$

4A $2x+1 = 9$ **5A** $2y-1 = 3$ **6A** $3m+5 = 8$

7A $8-x = 3$ **8A** $9-y = 0$ **9A** $6+k = 15$

10A $3m-5 = 7$ **11A** $5x+6 = 16$ **12A** $10 = 1+3y$

13A $3x+10 = 22$ **14A** $6-2y = 0$ **15A** $19 = 5k-1$

16B $26 = 3p-4$ **17B** $11x-11 = 33$ **18B** $21-7t = 0$

19B $6v+100 = 100$ **20B** $1+12w = 85$ **21B** $48-7x = 13$

LEARNING TO BALANCE

 This bag contains
8 metal weights.

● Each weight
weighs 1 kg.

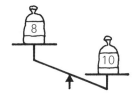 These bags are
not balanced.

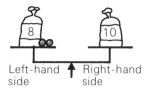

Left-hand ↑ Right-hand
side   side

Add 2 weights
to the left-hand
side. The bags
are balanced.

=== *Exercise 3* ===

Each of these can be balanced by *adding* weights to the left-hand side or to the right-hand side.
Explain how you could balance them.

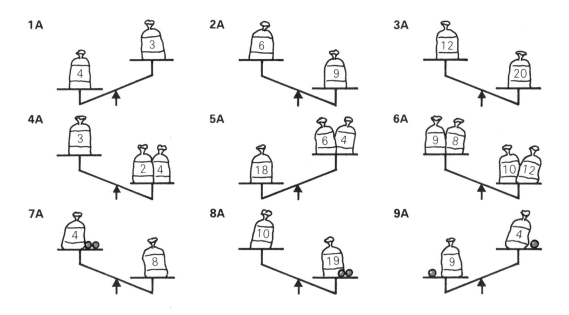

Explain how to balance these by *taking away* weights from the left- or right-hand sides.

There may be two different ways of balancing these. If there are, then explain both ways. You are not allowed to *remove* bags from the balance.

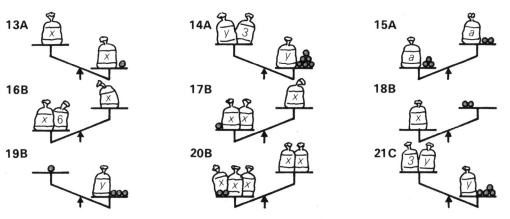

13A **14A** **15A** **16B** **17B** **18B** **19B** **20B** **21C**

22A For each picture in questions **13A–21C** write down the weight on each side of the balance. Put them in a table under the headings *Left-hand side* and *Right-hand side*.

EQUATIONS—1

Getting your balance

 This bag contains x metal weights.

● Each weighs 1 kg

Balance	Equation
Balanced.	$3x = x+6$ An equation.
Remove x weights from each side— the pans are still balanced.	$3x-x = x+6-x$ Subtract x from each side—we still have an equation.
Keep half of each side.	$2x = 6$ Divide each side by 2 (or 'cover up' x).
There must be 3 weights in the bag.	$x = 3$ The equation has been solved.

Aim

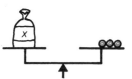

$$x \qquad = \qquad 3$$

| Finish with a single bag on the left-hand side of the balance. | Then count the weights on the right-hand side of the balance. | Finish with a single letter on the left-hand side of the equation. | Then the number the letter stands for is on the right-hand side of the equation. |

=== Exercise 4 ===

Write down equations for these sequences of pictures.

1A

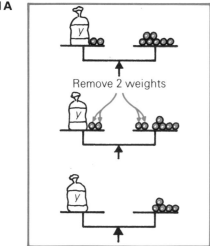

Remove 2 weights

2A

Remove 5 weights

3A

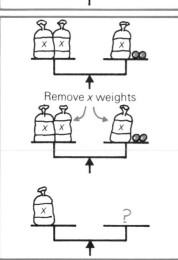

Remove *x* weights

4A

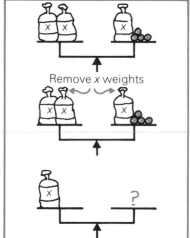

Remove *x* weights

5A

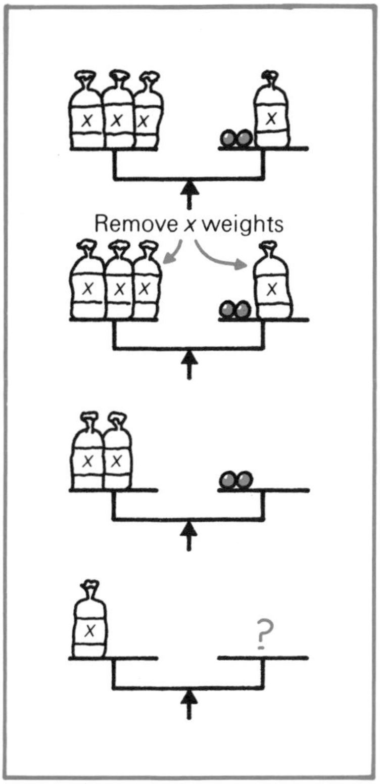
Remove x weights
?

6A

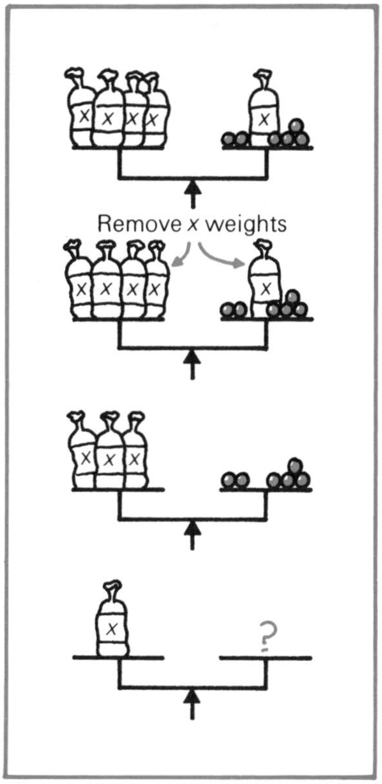
Remove x weights
?

7B

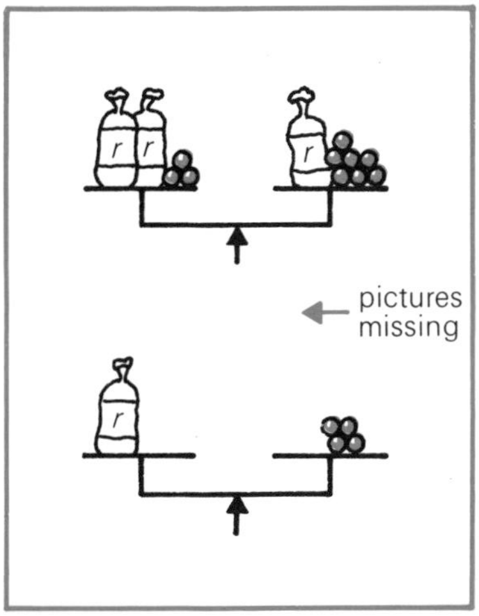
pictures
missing

8B

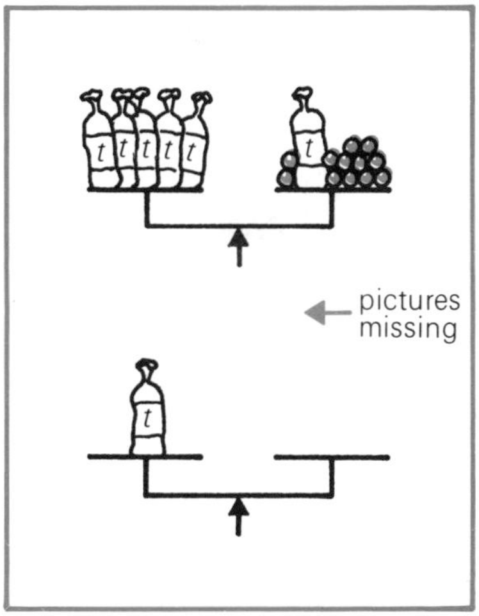
pictures
missing

9C

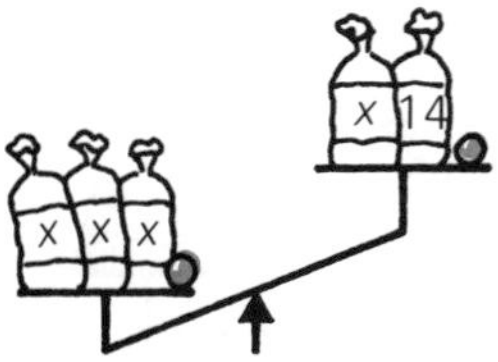
x
14

a

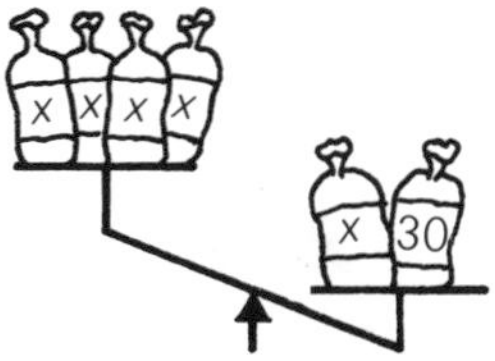
x
30

b

These bags just won't balance

How many weights are in each bag?

EQUATIONS—2

Keeping your balance

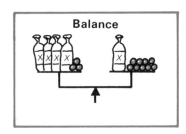

Equation

$$4x+3 = x+9$$
$$4x-x+3 = x-x+9 \qquad \text{Subtract } x \text{ from each side.}$$
$$3x+3 = 9$$
$$3x+3-3 = 9-3 \qquad \text{Subtract 3 from each side.}$$
$$3x = 6$$
$$x = 2$$

Each bag must contain 2 weights.

Check	**Balance**	**Equation**
Left side		$4x+3 = 4 \times 2+3 = 8+3 = 11$
Right side		$x+9 = 2+9 = 11$

Each side has 11 weights; they are balanced.   Each side = 11

─────────────────── *Exercise 5* ───────────────────

For each picture, make up an equation and then solve it.
Check your solution as shown above.

1A 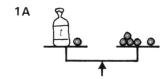   2A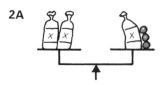

3A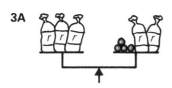

4A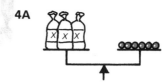

5A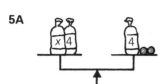

6A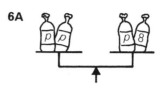

7B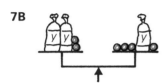

8B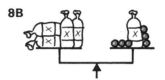

9C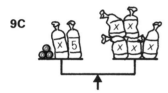

Losing your balance

10C Someone knocked these balances over. Each had been balanced.

a

b

In how many different ways might the bags and weights have been arranged on the balances? For each way that you can think of, write an equation and solve it, finding the number of weights in the bag to show that it is a sensible arrangement.

USING EQUATIONS

What can I afford?

EQUATIONS

I can afford

either 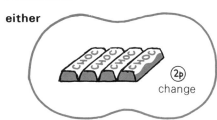   **or**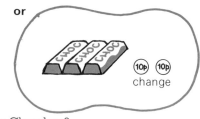

The Problem

They both cost the same. How much is a Choc bar?

Suppose costs x pence.

⊲ Introduce a letter

$$4x+2 = 3x+20$$
$$4x-3x+2 = 3x-3x+20$$
$$x+2 = 20$$
$$x+2-2 = 20-2$$
$$x = 18$$

⊲ Make an equation

⊲ Solve the equation

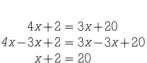 costs 18 pence.

⊲ What does the solution mean?

 is worth $4 \times 18 + 2 = 74$ pence

 is worth $3 \times 18 + 20 = 74$ pence.

⊲ Check the solution

Exercise 6

In each question you are shown two ways that I can spend all the money I have in my pocket. Make an equation from each pair of pictures. Then solve the equation, and check your solution.

1A

How much is a bag of chips?

2A

The fish costs 80p.

How much is a bag of chips?

3A

How much is a large cone?

4A

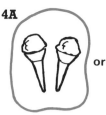

Flakes are 20p extra. How much is a small cone?

5A

 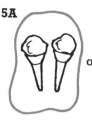 

How much is a cone at this shop, if flakes are still 20p?

6A

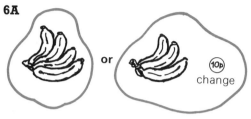

How much is a banana?

7A

How much is an envelope?

8B

How much is a bottle of Action Fizz?

9B

Pickled onions are 5p extra. How much is a bag of chips this time?

10B

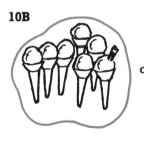 

Flakes are 15p extra here. How much is a cone now?

11C

Plastic bags are 6p extra. For 3 plastic bags you could buy 2 apples! How much is a banana?

Large envelopes are twice the price of small envelopes. What is the price of a small envelope?

12C

EQUATIONS—3

Off the balance

Balance		Equation
	3 weights have fallen out of one of the bags, but the bags are balanced.	$2x - 3 = x + 4$
	The 3 weights are put back in the bag. To keep the balance, 3 weights are added to the other side.	$2x - 3 + 3 = x + 4 + 3$
	All bags are full again, and still balanced.	$2x = x + 7$
	Remove 1 bag from each side.	$2x - x = x - x + 7$
	There must be 7 weights in the bag.	$x = 7$

=== *Exercise 7* ===

For each picture make up an equation, solve it and find the number of weights in each bag. Then check your solution with the picture, making sure the bags are balanced.

1A 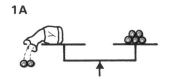 **2A** 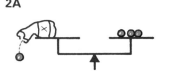 **3A**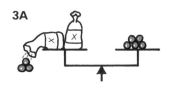

4A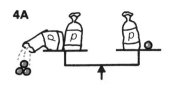

5A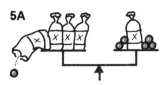

6B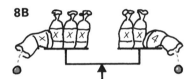

7B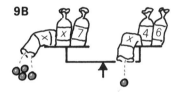

8B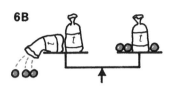

9B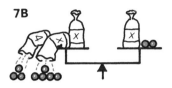

Mixed bags

10C

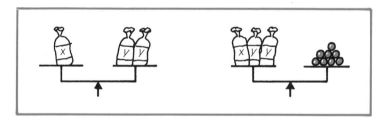

How many weights in the *x* and *y* bags?

SOLVING PROBLEMS

Bulk buying

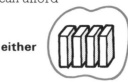 I can afford

either or   The Problem

Both cost the same.

How much was a packet?

 costs x pence. Introduce a letter

$4x - 10 = 3x$ Make an equation

$$4x - 10 + 10 = 3x + 10$$
$$4x = 3x + 10$$
$$4x - 3x = 3x - 3x + 10$$
$$x = 10$$

Solve the equation

 costs 10 pence. What does the solution mean?

 cost 40p less 10p discount. That's 30p. Check the solution

Cost 30p.

═══════════════ *Exercise 8* ═══════════════

In each question you are shown two things that I can afford. Both cost the same. Make an equation from each pair of pictures, and solve it.

1A Find the cost of a Choc bar.

2A Find the cost of a single.

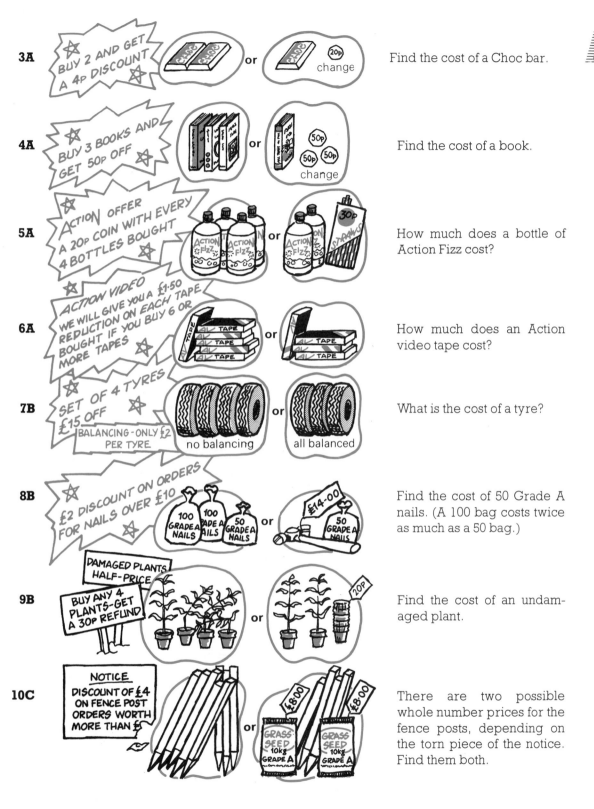

3A ☆ BUY 2 AND GET A 4p DISCOUNT ☆   or   (20p change)   Find the cost of a Choc bar.

4A ☆ BUY 3 BOOKS AND GET 50p OFF ☆   or   (50p 50p 50p change)   Find the cost of a book.

5A ☆ Action OFFER A 20p COIN WITH EVERY 4 BOTTLES BOUGHT ☆   or   (30p STRAWS)   How much does a bottle of Action Fizz cost?

6A ☆ ACTION VIDEO WE WILL GIVE YOU A £1.50 REDUCTION ON EACH TAPE BOUGHT IF YOU BUY 6 OR MORE TAPES ☆   or   How much does an Action video tape cost?

7B ☆ SET OF 4 TYRES £15 OFF ☆ BALANCING - ONLY £2 PER TYRE   no balancing   or   all balanced   What is the cost of a tyre?

8B ☆ £2 DISCOUNT ON ORDERS FOR NAILS OVER £10 ☆   (100 GRADE A NAILS, 100 GRADE A NAILS, 50 GRADE A NAILS)   or   (£14·00, 50 GRADE A NAILS)   Find the cost of 50 Grade A nails. (A 100 bag costs twice as much as a 50 bag.)

9B DAMAGED PLANTS HALF-PRICE / BUY ANY 4 PLANTS-GET A 30p REFUND   or   (20p)   Find the cost of an undamaged plant.

10C NOTICE DISCOUNT OF £4 ON FENCE POST ORDERS WORTH MORE THAN £   or   (£8·00 GRASS SEED 10kg GRADE A, £8·00 GRASS SEED 10kg GRADE A)   There are two possible whole number prices for the fence posts, depending on the torn piece of the notice. Find them both.

MAKING SURE

===== *Exercise 9* =====

Solve the equations:

1A $2x+1 = 7$	**2A** $3x+1 = 4$	**3A** $2x+3 = 9$
4A $2x = x+5$	**5A** $2a = a+1$	**6A** $3y = 2y+4$
7A $3b = b+6$	**8A** $4c = c+15$	**9A** $5d = 3d+10$
10A $2x-1 = x+6$	**11A** $2y+3 = y+4$	**12A** $3k-2 = k+10$
13A $5n+1 = 2n+7$	**14A** $6p-1 = 5p+1$	**15A** $4t+3 = t+6$
16B $3x-10 = 50$	**17B** $5y+10 = 100$	**18B** $10m-20 = 80$
19B $6x+3 = 2x+7$	**20B** $11x+1 = x+11$	**21B** $8y+5 = 3y+5$

Questions of balance

===== *Exercise 10* =====

1A

These two groups of luggage are the same weight.
What weight label should he put on the suitcases if they all weigh the same? What is the total weight of each group?

2A Do question **1A** again for 5 cases in the left-hand picture and 3 cases in the right-hand picture.

3A From this piece of wood
Jack can cut:

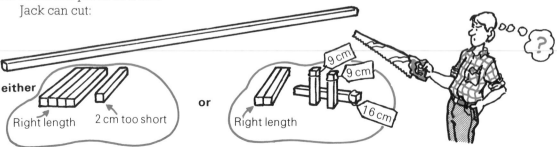

either

Right length   2 cm too short   or   Right length   9 cm
                                                                                           9 cm
                                                                                           16 cm

What is the 'right length'?   How long was the piece of wood?

4B

Half this tank can fill:

The other half can fill:

Each bucket is
1 litre short
of being full

How much does a bucket hold?

How much does the tank hold?

5C

I did 9 laps.

I only had 36 seconds of stops.

I did 7 laps.

I only had 1 minute 48 seconds
of stops.

The two drivers took the same total time, and averaged the same speed.
How long, on average, did 1 lap take? What was their total time?

1
The labels have fallen off these bags.
One of the bags contains more weights
than either of the other two, which both contain the same number of weights.

Explain how with just *one* weighing you can
be sure of finding the heavier bag.

2 Explain how it takes no more than *two*
weighings to be sure of finding the heavier
bag this time.

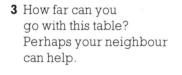

3 How far can you
go with this table?
Perhaps your neighbour
can help.

Number of bags	Minimum number of weighings, to be sure of finding the heavier bag
2	. . .
3	. . .
4	. . .
.	. . .
.	. . .
.	. . .

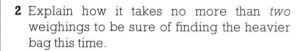

CHECK-UP ON **EQUATIONS**

1A

Copy and complete:

This weighing machine is _____.
$x+3 = 6$ is the _____ for the picture.

$x = 3$ is the _____. Three weights in the bag
gives six weights on each side. The solution has now been _____.

2A

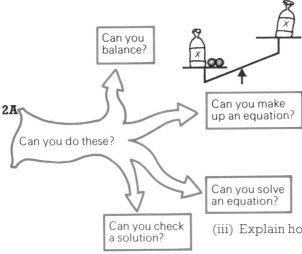

Can you balance?

Can you do these?

Can you make up an equation?

Can you solve an equation?

Can you check a solution?

a Explain how to balance this weighing machine.

b (i) Make up an equation for this picture:

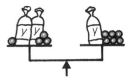

(ii) Solve the equation above.

(iii) Explain how to check your solution.

3B

Can you solve this problem?

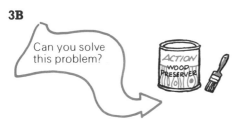

There is enough in this tin to treat

either

with enough left for 1 post.

or

with enough left for 11 posts.

How many posts are there in a bundle?

4C

Can you solve this problem?

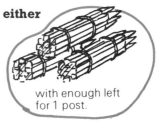

How many balls of wool do you have?

Nearly enough for four pullovers.

How many more would you need for four?

Three balls.

I only want one pullover.

Then I'll have 33 balls of wool left over.

How many balls of wool are needed for a pullover?

LENGTH

Class discussion—Units of length through the ages

Many different units of length have been used in the past. Here are some of them.

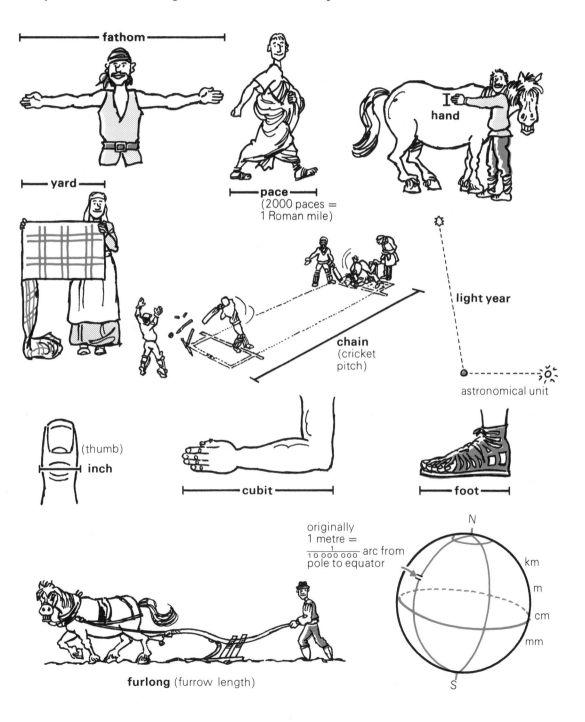

fathom

yard

pace
(2000 paces =
1 Roman mile)

hand

chain
(cricket
pitch)

light year

astronomical unit

(thumb)
inch

cubit

foot

originally
1 metre =
$\frac{1}{10\,000\,000}$ arc from
pole to equator

N

km

m

cm

mm

S

furlong (furrow length)

Units of length

The **metre** (m) is now the standard unit of length in many countries.

1 metre = 1 650 764 times the wavelength of the orange red line of the krypton gas spectrum.

The **metric system** of measures is based on tens, the same as our decimal number system, so calculations are simpler than with other systems.

1 metre is divided into 100 **centimetres** (cm).

1 centimetre is divided into 10 **millimetres** (mm).

A length of 1000 metres is called a **kilometre** (km).

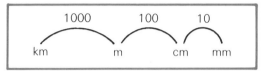

―――――――――――――――――――― *Exercise 1* ――――――――――――――――――――

1 a Use your ruler to measure the widths of 1p, 2p, 5p and 10p coins:
   (i) in centimetres, to the nearest 0·1 cm   (ii) in millimetres.
   b Which unit would you use to measure the thickness of a coin?

2 Measure the length, breadth and thickness of this book in centimetres, to the nearest 0·1 cm, and also in millimetres.

3 Which units—mm, cm, m or km—would you use to measure:
   a your height   **b** the height of the school
   c the length of a pencil   **d** the length of Britain (Land's End to John O'Groats)
   e the thickness of a book   **f** the thickness of a fingernail
   g the distance to the school office **h** the distance to New York?

4 Estimate the lengths or distances in question **3**.

5 a Without using your ruler, draw a line 'freehand' which you think is 3 cm long.
   b Measure its length with a ruler.
   c Write down the difference in lengths. Remember to include the unit of length.

6 Repeat question **5** for lines of lengths:
   a 5 cm   **b** 8 cm   **c** 5 mm   **d** 12 mm

―――――――――――――――――――― *Exercise 2A* ――――――――――――――――――――

1 Alistair, Ian and Hassan took part in the triple jump event at the school sports.
   a The three parts of Alistair's jump measured 1·3 m, 1·3 m and 1·5 m. What total length did he cover?
   b Ian's results were 1·2 m, 1·3 m, 1·4 m. Hassan managed 1·4 m, 1·4 m, 1·2 m. Work out the total length of each of their jumps.
   c Who won?

2 Six books were put in a pile. The thicknesses of the books were 14 mm, 20 mm, 31 mm, 24 mm, 13 mm and 25 mm. What was the height of the pile?

3 Nathan's new pencil was 17 cm long. After a month it was only 14·3 cm long. What length of pencil had been used?

4 At the start of a journey a car's mileage recorder read 126 574 km. After the journey it read 126 611 km. How long was the journey?

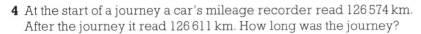

5

Kath buys a strip of seven 5p stamps. The width of one stamp is 21 mm. What is the length of the strip?

6 Dawn's small sister has a 'concertina' book like this. Each page shows one letter of the alphabet, and is 8·3 cm wide. How long is the book when opened out flat?

7 Outside a car factory new cars are parked bumper to bumper. A line of six cars takes up a space 28·8 m long. How long is a car?

8 Not counting the covers, the thickness of a book is 30 mm. The book has 600 pages. How thick is each sheet of paper in the book?
(*Careful!* There's a catch in this question.)

=========================== *Exercise 2B* ===========================

1 A botany class was doing a study of plants and flowers. Keith measured some foxglove plants. Their heights were:
56 cm, 57 cm, 61 cm, 62 cm, 60 cm and 64 cm.
  a Add up the heights.
  b Divide the total by the number of plants.
  c Write your answer: Average height of foxgloves = . . . cm.

2 Diana decided to measure some primroses. Here are their heights:
5 cm, 6 cm, 7 cm, 4 cm, 4 cm, 3 cm, 6 cm, 5 cm, 3 cm and 7 cm.
Calculate the average height of these primroses.

3 Some books on wild flowers describe the heights of plants like this:

Tall	60 cm and over
Medium	30 cm and less than 60 cm
Short	10 cm and less than 30 cm
Low	Less than 10 cm

Use this system, and Keith and Diana's results, to describe:
a the foxglove   **b** the primrose.

4 Copy and complete this table of some other plants studied by the botany class.

Name of plant	Heights (in cm)						Average height (to nearest 0·1 cm)	Description
Dandelion	13	12	11	10	13	13		Short
Daisy	5	5	6	4	4	5		
Poppy	35	37	35	36	36	37		
Buttercup (meadow)	27	33	29	31	28	27		

===== *Exercise 2C* =====

1 Jill's job in her youth club's project on the weather is to measure the rainfall. To do this she collects the rain in a jar. Each day she measures the depth of water in the jar, and every Saturday night she empties the jar.

a Copy and complete this table.

Day	Sunday	Monday	Tuesday	Wednesday	Thursday	Friday	Saturday
Reading	7 mm	15 mm	20 mm	26 mm	32 mm	44 mm	44 mm
Rainfall	7 mm	8 mm					

b Calculate the average daily rainfall, to the nearest millimetre, for the week shown in the table.
c Which days had rainfall above the average for the week?
d What was the weather like between Friday and Saturday?

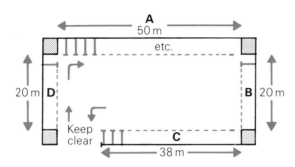

LENGTH, AREA AND VOLUME

2 A planner is working on the layout for a car park. He has to allow a width of 2·5 metres for each car.
  a Section A is 50 metres long. How many spaces can the planner mark out?
  b How many cars will the car park hold altogether in sections A, B, C and D?
  c Why are the corners not used for parking?
  d Design your own car park for a rectangle of ground 45 m long and 30 m broad, allowing 3 m for each car width.

===== *Exercise 3A* =====

1 The Wilson family is on holiday at Coniston, in the Lake District. On Tuesday they take a trip to Windermere. They go by Ambleside, and return by Staveley.
  a What is the distance going to Windermere?
  b What is the distance on the return journey?
  c How much longer is the return journey? On Thursday they set off for Morecambe by the shortest route.
  d What towns will they pass through?
  e They return the same day. How far did they travel, there and back?

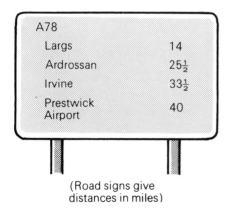

(Road signs give distances in miles)

2 Linda Fleming travels from Greenock to Prestwick Airport. She sees this sign as she leaves Greenock.
  a Draw a line to represent her route from Greenock to Prestwick Airport. Mark the towns shown on the road-sign, and their distances from Greenock.
  b How far is she from Prestwick Airport as as she passes the sign?
  c How far is she from the airport when she reaches Largs?
  d How far apart are Ardrossan and Irvine?
  e Troon is between Irvine and Prestwick. It is 5 miles from Irvine. Redraw the road sign to include Troon.

LENGTH, AREA AND VOLUME

1 The cost of placing an advert in a newspaper depends on what is being advertised, and the length of the advertisement. The table below shows some of the charges.

Category	Cost per centimetre
Employment	£9·80
Property	£8·40
Entertainment	£8·25
Motor cars	£7·50
Charities	£3·00

Calculate the cost of placing each of these four ads.

a

FOR SALE ☆
★
USED CARS
012 369 1215

3 cm

b

☆
PICTURE HOUSE
EL CID
Programme starts:
2.30pm 5.30pm 8.30pm

6 cm

c

WANTED
CHEF
Contact:
HOTEL SPLENDIDE
012 468 1012

5 cm

d

TO LET
OFFICE SPACE
J. BLOGGS & Co
2 HIGH STREET
NEWTOWN
Phone 012 345 6789

7 cm

420   480
810
480
1000   1200
780

All measurements are in millimetres

2 'KIRSTY' dolls are made one-sixth of life-size.
Furniture for KIRSTY dolls will also be one-sixth life-size.
The picture shows the sizes of an actual table and chair.
The back of a chair is 810 mm high. So the KIRSTY doll's chair will have a back which is $810 \div 6 = 135$ mm high.
  a Scale down the other sizes given for the chair in the same way.
  b Calculate, to the nearest millimetre, the length, breadth and height of the doll's table.
  c Draw a picture of the doll's table and chair, and mark in their measurements.

Frank Wylie wants to build a fence round his garden.
He needs palings, backing strips and posts.

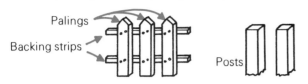

Palings

Backing strips

Posts

1 This is a drawing of his rectangular garden. Calculate the total length of fencing needed.
(He will put in a gate later.)

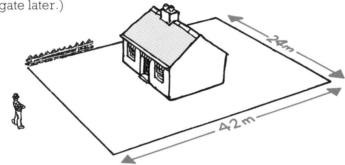

24 m

42 m

2 Sections of fencing are made up like this:

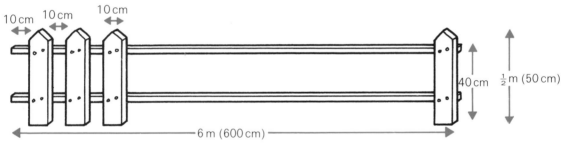

10 cm   10 cm   10 cm

40 cm   ½ m (50 cm)

6 m (600 cm)

How many sections will he need?

3 To make up one section, calculate:
  a the total length of backing strip
  b the number of pieces of paling
  c the total length of paling.

4 The fence is held up by posts. These are sunk into the ground to a depth of 25 cm, as you can
see in the drawing below. How long is each post?

5 Frank needs two posts for each fence
section and one for each corner. How
many posts does he need? What is the
total length of 'post' wood needed?

6 Write down an estimate for the lengths
of paling, backing strip and post needed
for the whole fence.

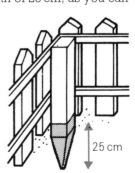

25 cm

AREA

Class discussion

A man is adding a room to his house. He goes to a DIY superstore to find out the sizes and shapes of window frames.

He sees two that he likes, and wonders which would let in more light.

The actual unit of length does not matter at present.

a Which window frame is broader?
b Which is taller?
c Which has the greater perimeter?
But which window lets in more light?
Here are two rectangles on squared paper to stand for the windows.

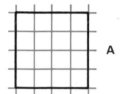

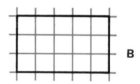

d Which rectangle has more squares in it?
e Which window uses more glass?
f Which window lets in more light?

> The number of squares gives a measure of the **area** of each window, or the amount of surface of each window.
> The area of window A is 16 square units.

Washing an elephant takes longer than washing a dog because an elephant's skin has a greater *area* than a dog's skin.

LENGTH, AREA AND VOLUME

1 Find the areas of these shapes by counting squares. For the curved shapes, half a square or more counts as 1 square. Don't count smaller parts.

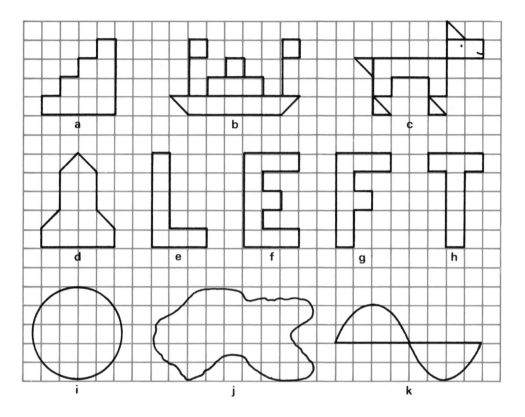

2 Find the areas of these windows by drawing them on squared paper, and counting squares.

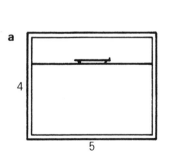

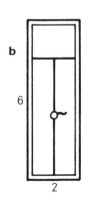

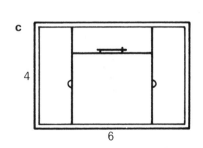

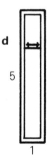

LENGTH, AREA AND VOLUME

3 Counting squares is all right when the numbers are small, but what about Mr Steel's problem? The roof of his house needs to be retiled.

It is not easy to count the tiles one by one, so he looks for a better way.

a How many rows of tiles are there?

b How many tiles are there in each row?

c So how many tiles are there on one side of the roof?

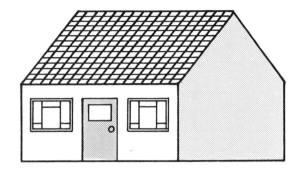

4 A home computer makes a grid on the television screen so that it can print words on the screen.

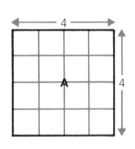

The grid has 25 rows of squares, with 40 squares in each row.

a How many squares make up the grid?

b Computer books describe this as a 25×40 grid. Explain why.

You can now use this method to calculate the areas of windows A and B.

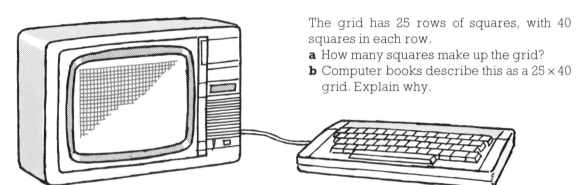

Area of window A
= 4×4 square units
= 16 square units.

5 Calculate the area of window B in this way.

6 Calculate the areas of the windows in question **2** in this way.

To calculate an area:

a Count squares.
b For a rectangle, multiply its length by its breadth.
A **formula** is $A = l \times b$, or $A = lb$, where A is the area, and l and b the length and breadth, all in suitable units.

Units of area

I want to know the area of this Airmail Sticker.

(i) I measure its length and breadth in millimetres.

20 mm

50 mm

Breaking it up into squares I have:

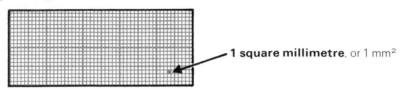

1 square millimetre, or 1 mm²

Altogether there are $50 \times 20 = 1000$ squares.
The lengths are measured in millimetres, so the
area of the sticker is 1000 **square millimetres**, or 1000 mm².

(ii) I measure its length and breadth in centimetres.

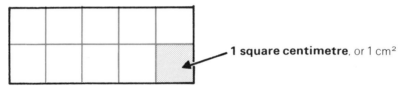

1 square centimetre, or 1 cm²

This time there are $5 \times 2 = 10$ squares.
Since the lengths are measured in centimetres, the
area of the sticker is 10 **square centimetres**, or 10 cm².

(iii) A larger unit of area is the **square metre** (m²).

Exercise 5

Find the area of each object in questions **1–10**. Remember to include the unit of area with your answer.

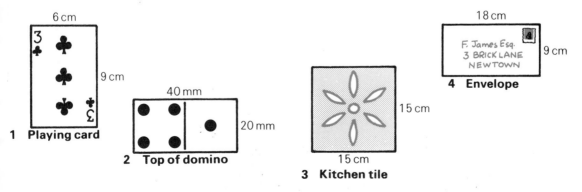

6 cm

3 ♣
♣
♣
♣ ♣
9 cm

1 Playing card

40 mm

20 mm

2 Top of domino

15 cm

15 cm

3 Kitchen tile

18 cm

F. James Esq.
3 BRICK LANE
NEWTOWN
9 cm

4 Envelope

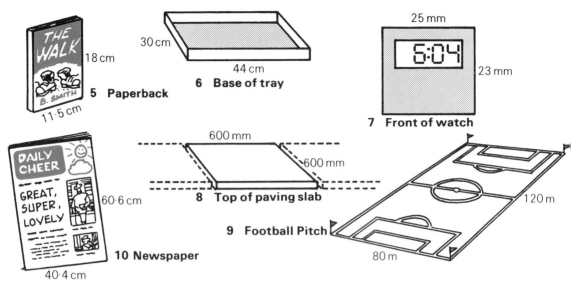

5 Paperback

THE WALK
B. SMITH
18 cm
11·5 cm

30 cm
44 cm
6 Base of tray

25 mm
6·04
23 mm
7 Front of watch

600 mm
600 mm
8 Top of paving slab

DAILY CHEER
GREAT, SUPER, LOVELY
60·6 cm

10 Newspaper
40·4 cm

9 Football Pitch
120 m
80 m

11 Sketch a rectangle which has an area of 12 cm². Mark its length and breadth. Can you draw two more rectangles which have areas of 12 cm²? Can you draw two more? How many do you think there are?

$$=========== \textit{Exercise 6A} ===========$$

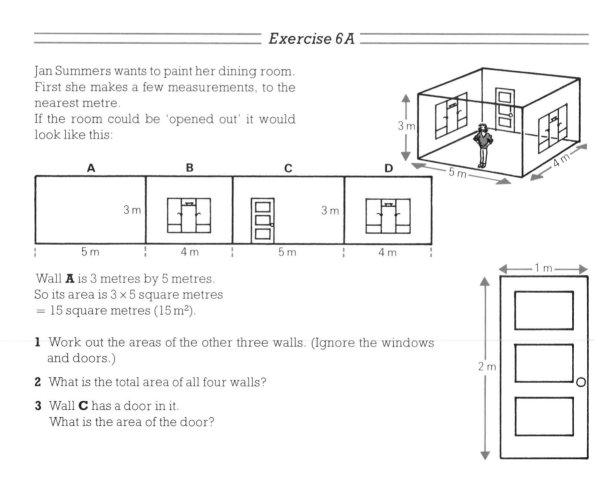

Jan Summers wants to paint her dining room.
First she makes a few measurements, to the
nearest metre.
If the room could be 'opened out' it would
look like this:

3 m
5 m
4 m

A	B	C	D
3 m			3 m
5 m	4 m	5 m	4 m

Wall **A** is 3 metres by 5 metres.
So its area is 3 × 5 square metres
= 15 square metres (15 m²).

1 Work out the areas of the other three walls. (Ignore the windows
and doors.)

2 What is the total area of all four walls?

3 Wall **C** has a door in it.
What is the area of the door?

1 m
2 m

4 Walls **B** and **D** both have a window like this.
What is the area of each window?

5 The door and the windows are not to be painted.
What is the total area that will be painted?

6 How many tins of paint like the one shown must she buy?

7 Jan decides to paint the ceiling as well, using the same paint.
  a Calculate the area of the ceiling.
  b Will she have to buy more paint? Explain your answer.

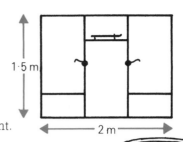

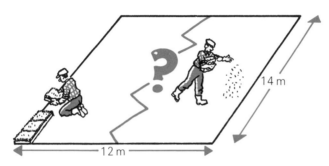

=== *Exercise 6B* ===

Mr Green is preparing some ground to make a new lawn. He can either lay turf or sow grass seed.

A turf is a $\frac{1}{2}$ m by $\frac{1}{2}$ m square of grass like the one shown here.

1 What is the area of one turf?

2 How many pieces of turf does he need to cover one square metre?

3 The lawn will be a 12 metre by 14 metre rectangle. What is its area?

4 How many pieces of turf does he need to buy?

5 If each turf costs 50p, what will it cost to lay his lawn?

Grass seed is sold in boxes which contain enough seed for $10 \, m^2$. Each box of seed costs £2·50.

6 How many boxes of seed would Mr Green need?

7 What would the cost be?

8 Which method is cheaper?

9 Why might someone choose the dearer method?

Exercise 6C

1 Nina is making place-mats for the dinner table.
She takes strips of material like this:

and weaves them like this:

into a square-shaped mat like this:

a What is the area of one strip of the material?
b How many strips does she use in making the mat?
c What is the total area of all the strips used?
d What is the area of the top of the finished mat?
e Should the answers for **c** and **d** be the same? Explain.
f Nina decides to make a large square mat for the centre of the table. She thinks it should have an area of 2500 cm².
How many 50 cm by 2 cm strips of material will she need?

A closed loop of rope has a length of 16 metres. Bryan has been given 4 pegs and this rope.

His job is to peg out the largest rectangle he can; that is, the rectangle with the largest area.

Copy and complete this table:

Length (m)	1	2	3	4	5	6	7
Breadth (m)	7	6	5				
Area (m²)	7	12					

Describe the shape Bryan should mark out.
Describe the lengths of the sides and the area of the largest possible rectangle if the loop of rope has a length of:
a 25 metres **b** 36 metres **c** 49 metres **d** 30 metres **e** *x* metres

VOLUME

Class discussion

Here are two boxes made to hold sugar cubes. Their measurements are in the same units. Which box is larger?

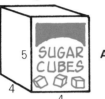   A      B

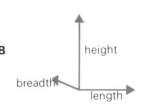

LENGTH, AREA AND VOLUME

1 Which box is: **a** longer   **b** broader   **c** taller?

2 Which has the larger face?

3 Which has the larger base?

But which box holds more sugar cubes?

To find out, we will fill each box with sugar cubes like this.

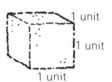

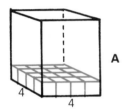   A   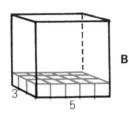   B

4 For box A:
  a Put in one layer. How many cubes?
  b How many layers will it hold?
  c How many cubes will it hold?

5 Repeat question **4** for box B.

6 a Which box holds more cubes?
  b So which box is larger?

The number of cubes gives a measure of the **volume** of each box, or the amount of space in each box.

This cube has a volume of 1 cubic centimetre, or 1 cm^3.

If the measurements were in centimetres, then:
      The volume of box A
      $= 4 \times 4 \times 5$ cubic centimetres
      $= 4 \times 4 \times 5$ cm^3
      $= 80$ cm^3

Copy and complete these calculations.

1

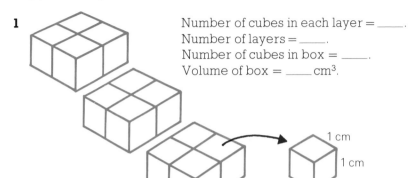

Number of cubes in each layer = _____.
Number of layers = _____.
Number of cubes in box = _____.
Volume of box = _____ cm³.

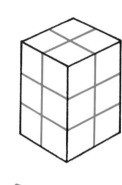

2

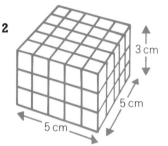

Number of cubes in each layer = _____.
Number of layers = _____.
Number of cubes in box = _____.
Volume of box = _____ cm³.

3

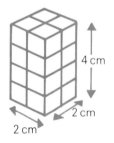

Number of cubes in each layer = _____.
Number of layers = _____.
Number of cubes in box = _____.
Volume of box = _____ cm³.

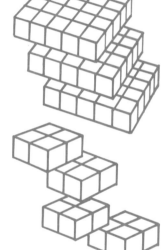

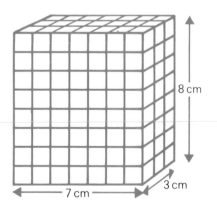

5

4

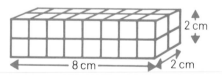

Volume of box = 8 × _____ × _____ = _____ cm³.

Volume of box = _____ × _____ × _____ = _____ cm³.

> **To calculate the volume of a box (cuboid)**, multiply its length by its breadth by its height.
> A **formula** is $V = l \times b \times h$, or $V = lbh$, where V is the volume, and l, b, h are the length, breadth and height, all in suitable units.

Units of volume

The units of volume include the cubic millimetre (mm^3),
the cubic centimetre(cm^3),
and the cubic metre (m^3).
Liquids are often measured in **litres**; 1 litre = $1000 \, cm^3$.

=========== *Exercise 8* ===========

Find the volume of each box. All measurements are in centimetres. Remember to give units in your answers.

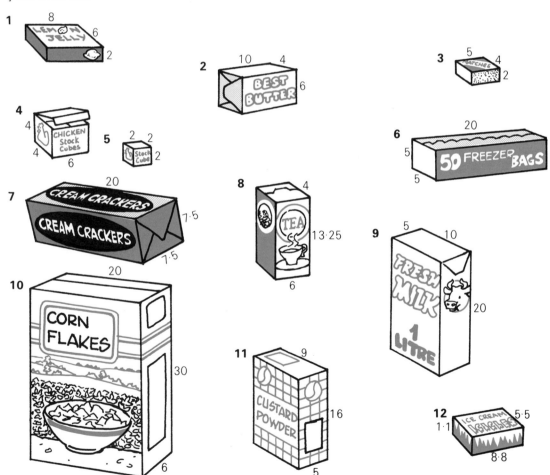

1 8, 6, 2

2 10, 4, 6

3 5, 4, 2

4 4, 4, 6

5 2, 2, 2

6 20, 5, 5

7 20, 7·5, 7·5

8 4, 13·25, 6

9 5, 10, 20

10 20, 30, 6

11 9, 16, 5

12 5·5, 1·1, 8·8

13 Why are so many packets in the shape of cubes and cuboids?

When fridges and freezers are advertised we are often told their **cubic capacity**. This means the amount of space inside them. The larger the capacity, the more food they will hold.

1 Look at this chest freezer. Its inside is a box like this:

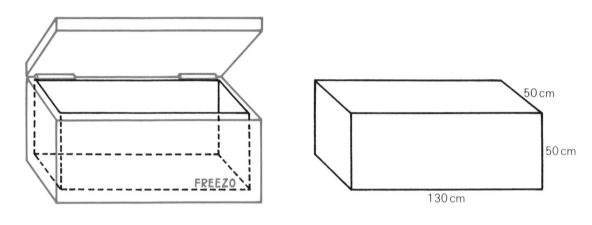

What is the inside volume, or capacity, of the freezer?

2 This is a diagram of a fridge/freezer.
It has two separate compartments.

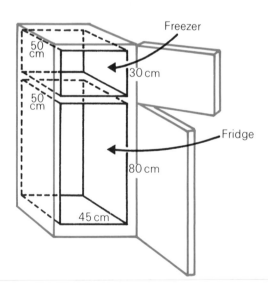

Calculate the capacity of:
a the freezer   **b** the fridge   **c** the fridge/freezer.
Remember to include units in your answer.

3 You will have noticed that all your answers are quite large.
For this reason, shops don't give the capacity of a fridge in cubic centimetres. A larger unit, the litre, is used. 1 litre = 1000 cm³.
Divide all your answers for questions **1** and **2** by 1000 to turn them into litres.

LENGTH, AREA AND VOLUME

The size of a central heating radiator for a room depends on the volume of the room.

Radiator	Volume of room
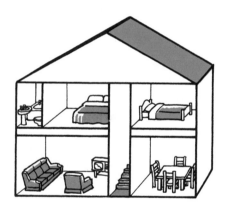 size 1	Less than 65 m³
size 2	From 65 m³ to 99 m³
size 3	More than 99 m³

The dining room in the house below has a height of 4 metres. Its length is 5 metres and its breadth is also 5 metres.
So its volume $= 4 \times 5 \times 5 \, \text{m}^3 = 100 \, \text{m}^3$.
From the chart, you can see that a size 3 radiator is the best one for this dining room.

Copy and complete this table. The lengths given are in metres.

	Room	Length	Breadth	Height	Volume	Radiator size
1	Main bedroom	5	5	3		
2	Small bedroom	5	4	3		
3	Bathroom	4	4	3		
4	Kitchen	4	4	4		
5	Lounge	6	5	4		
6	Hall (and stairs)	5	2	7		

LENGTH, AREA AND VOLUME

Look at these two cartons of juice.

Check that their volumes are equal.

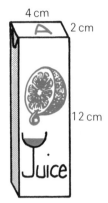

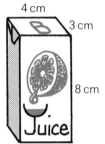

They both hold the same amount of juice; but the area of special cardboard needed to make each of them is not the same. Copy and complete this table.

	Carton A		Carton B	
Face	$l \times b$ (cm²)	area (cm²)	$l \times b$ (cm²)	area (cm²)
Left	2×12	24		
Right	2×12	24		
Top	2×4	8		
Bottom	2×4	8		
Front	4×12	48		
Back	4×12	48		
Total area of card		160		

Do you find that carton B needs less cardboard than carton A?
It would probably be cheaper to make.

Find another carton with the same volume, which needs even less cardboard than carton B.

Can you design the carton which will use the least amount of cardboard to hold 216 cm³ of juice?

Try to do the same for a carton holding 343 cm³, and one holding 512 cm³.

206

1A, B Length

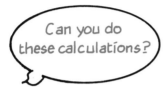

Do you know the units of length?

Can you do these calculations?

Investigate.

a What is the standard unit of length in the metric system?

b Name an everyday object which is about:
   (i) 1 metre long   (ii) 1 centimetre long
   (iii) 1 millimetre long.

c Write down the short form of each unit in question **b**; for example, 1 kilometre = 1 km.

d A piece of elastic is 93 cm long. How much further would you have to stretch it to make it 1 m long?

e (i) 1000 cards make a pile 1 m high. Calculate the thickness of 1 card in mm.
   (ii) How many cards would make a pile 1 cm high?

f (i) A charity is being given all the money in 1 km of one-pence pieces, placed side by side in a straight line. 1 penny is 2 cm wide.
   How much money will the charity get?
   (ii) Investigate this idea for different coins.

2A, B, C Area

Do you know the units of area?

Can you do these calculations?

a Write down the area of a square with side 1 millimetre long.

b Write down the short form of each of these units for measuring area:
   (i) 1 square millimetre
   (ii) 1 square centimetre
   (iii) 1 square metre.

c What unit would you use for the area of:
   (i) the classroom floor
   (ii) a postage stamp
   (iii) your desk top?

d $A = lb$. Write down the meaning of this formula in words.

e Calculate the area of each of these:

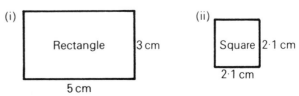

(i) Rectangle 3 cm / 5 cm
(ii) Square 2·1 cm / 2·1 cm

LENGTH, AREA AND VOLUME

f A painting is 24 cm long and 18 cm broad. It has a border 5 cm wide all round it. Calculate the area of the outside rectangular shape.

g A carpet has an area of 36 m². Its length is 9 m. Calculate its breadth.

h Investigate how many mm² make up 1 cm².

3A, B, C   Volume

a What is the volume of a cube with edge 1 cm long?

b Write down the short form of each of these units for measuring volume:
  (i) 1 cubic millimetre
  (ii) 1 cubic centimetre
  (iii) 1 cubic metre.

c What unit would you use for the volume of:
  (i) a box of cereal   (ii) a room
  (iii) a matchstick?

d $V = lbh$. Write down the meaning of this formula in words.

e Calculate the volume of each of these boxes:

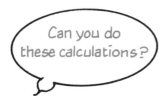

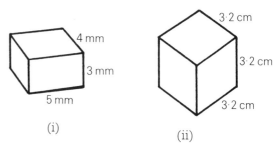

(i)   (ii)

f A carton of juice is 10 cm by 5 cm by 20 cm.
  (i) Calculate its volume, in cm³.
  (ii) What other name is given to this quantity?

g A box has a volume of 1000 cm³. The area of its base is 25 cm². Calculate the height of the box.

h Investigate how many cm³ make up 1 m³.

TYPES OF ANGLE

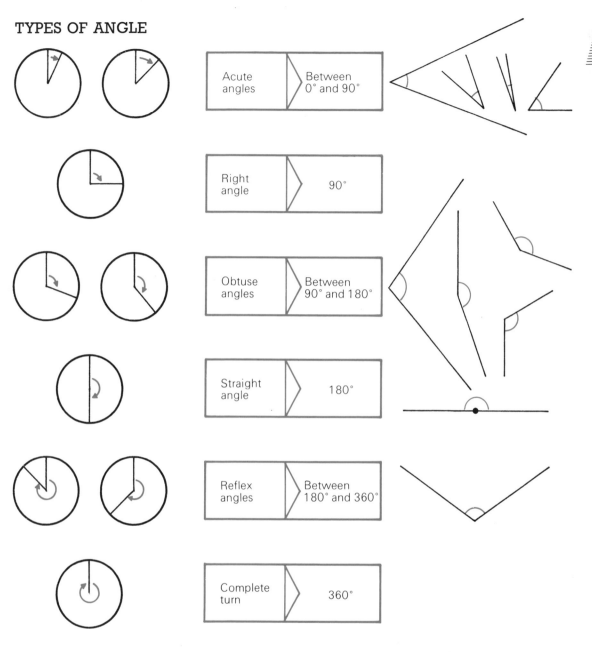

Acute angles	Between 0° and 90°
Right angle	90°
Obtuse angles	Between 90° and 180°
Straight angle	180°
Reflex angles	Between 180° and 360°
Complete turn	360°

Able to label

Exercise 1

1A Use your protractor to check that this angle measures 28°.
Since 28° is between 0° and 90° the angle is acute.
The angle has been labelled to show this.

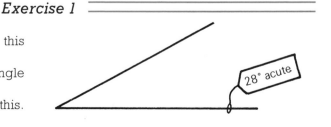

28° acute

2A Copy these labels into your notebook, and fill in the size and type of each angle. You will need to use your protractor.

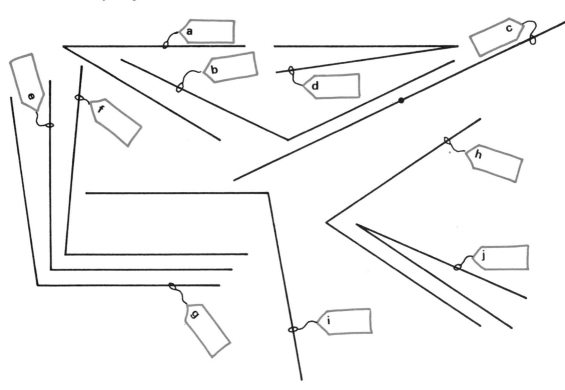

3A Here is a collection of labels. Some of the labels are wrong. Find them.

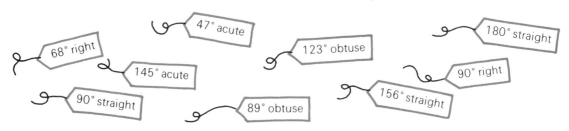

68° right

47° acute

145° acute

123° obtuse

180° straight

90° right

90° straight

89° obtuse

156° straight

4A Draw angles for the labels that are correct in question **3A**.

5A Make up a table and list the names of all the acute angles, right angles and obtuse angles in these shapes. For example, ∠BAC.

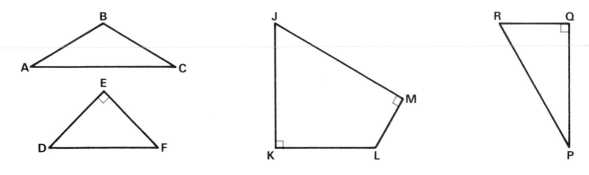

Possibilities

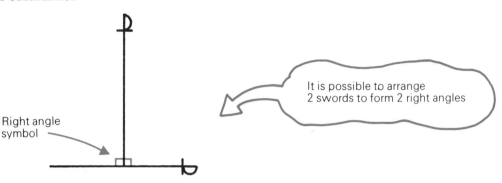

===== *Exercise 2* =====

Draw diagrams to show how it is possible to arrange two swords to form:

1A An acute angle   **2A** An obtuse angle   **3A** A right angle

4A Four right angles   **5A** One acute angle and one obtuse angle

6A Two acute angles and two obtuse angles.

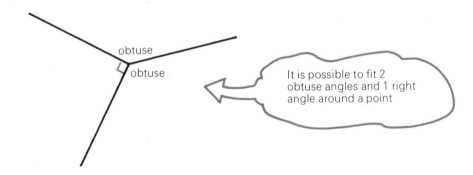

In this exercise, if your answer to a question is **yes**, draw a diagram to show how it is possible. If your answer is **no**, explain why not.
Is it possible to fit the following?

7A Two acute angles into a right angle   **8A** Two acute angles into an obtuse angle

9A Four right angles around a point   **10A** Three obtuse angles around a point

11B Two obtuse angles into an obtuse angle   **12B** Four obtuse angles around a point

13B An acute and an obtuse angle into a reflex angle.   **14B** Two reflex angles around a point

Is it possible for a triangle to have the following?

15B Three acute angles   **16B** Two acute angles and one obtuse angle

17B One acute angle and two obtuse angles   **18B** Three obtuse angles

Can you draw these lines?

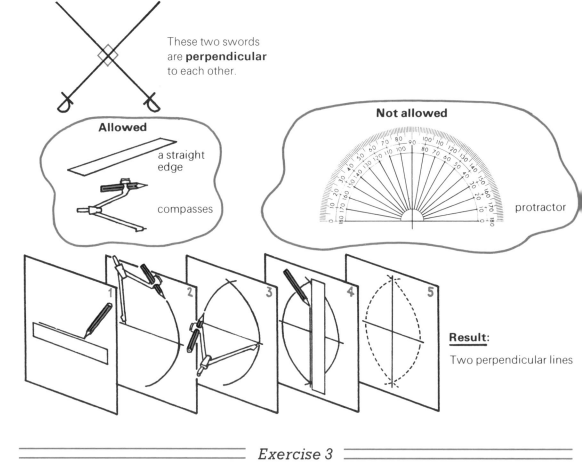

These two swords are **perpendicular** to each other.

Allowed

a straight edge

compasses

Not allowed

protractor

1 2 3 4 5

Result:

Two perpendicular lines

Exercise 3

Copy these diagrams, and draw the perpendicular lines described.

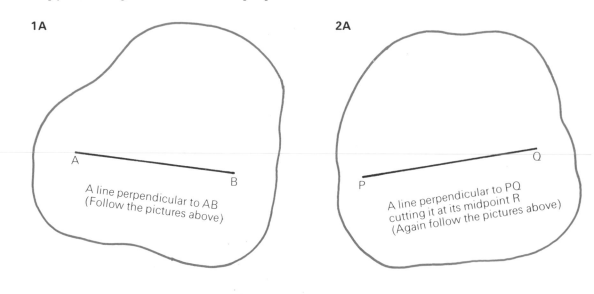

1A

A ———————— B

A line perpendicular to AB
(Follow the pictures above)

2A

P ———————— Q

A line perpendicular to PQ
cutting it at its midpoint R
(Again follow the pictures above)

3A

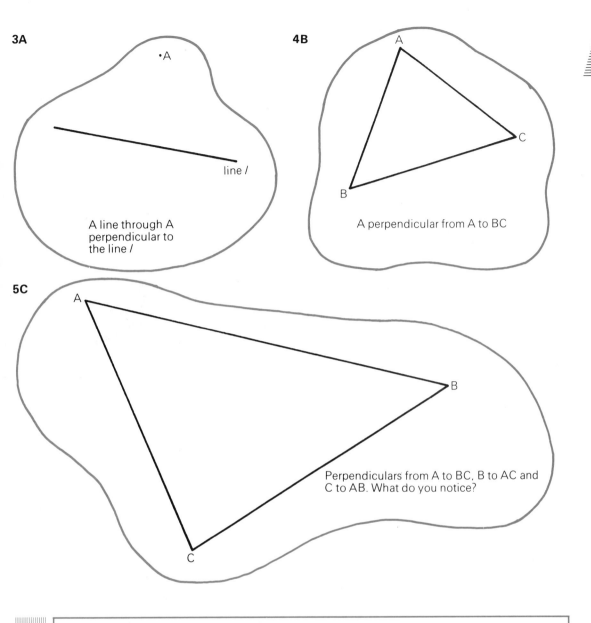

•A

line *l*

A line through A
perpendicular to
the line *l*

4B

A

C

B

A perpendicular from A to BC

5C

A

B

C

Perpendiculars from A to BC, B to AC and
C to AB. What do you notice?

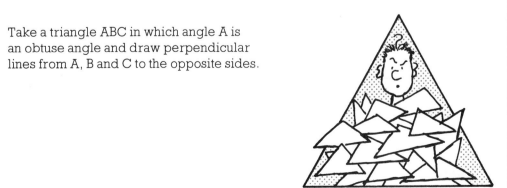

Take a triangle ABC in which angle A is
an obtuse angle and draw perpendicular
lines from A, B and C to the opposite sides.

HORIZONTAL AND VERTICAL

Class discussion

1 Here is a marble on a board.
The board is **horizontal**.
Will the marble roll?

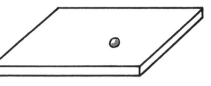

2 The balanced pencil is **vertical**.
How many horizontal pencils are there?
Why is the marble not rolling?
If the marble starts to roll, what has happened to the board? Will the balanced pencil still be vertical?

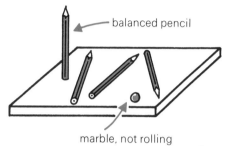

balanced pencil

marble, not rolling

3 How many pencils are vertical and how many are horizontal.
If the horizontal pencils are perpendicular to each other, how many right angles do the three pencils form?

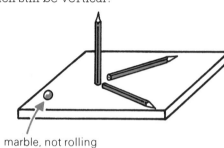

marble, not rolling

4 a Describe any horizontal lines and surfaces you can see around you.
b Now look for vertical lines and surfaces.

Exercise 4

1A Sketch the following. Then go over horizontal lines in one colour, and vertical lines in another colour.

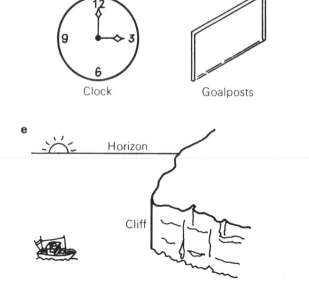

a Clock b Goalposts c Brick wall d Cricket stumps and bails

e Horizon / Cliff

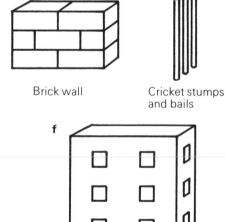

f Block of flats

ANGLES AGAIN

214

2A Which of these are horizontal, and which are vertical?

a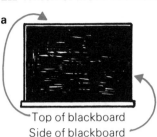
Top of blackboard
Side of blackboard

b
Coin standing
on edge

c
Surface of a
glass of milk

d
Wire from ceiling
holding a light

3A Which of these statements are true and which are false?

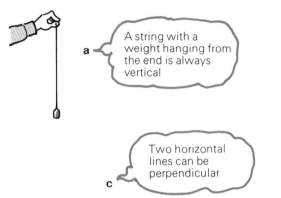

a A string with a weight hanging from the end is always vertical

b A horizontal line is never vertical

c Two horizontal lines can be perpendicular

d The horizon is horizontal

4B Which of these statements are true and which are false?

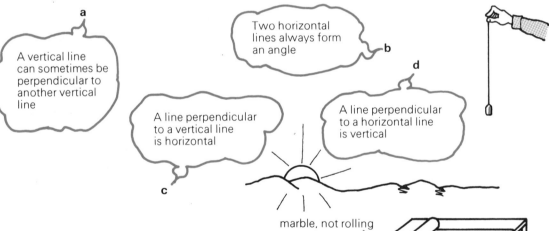

a A vertical line can sometimes be perpendicular to another vertical line

b Two horizontal lines always form an angle

c A line perpendicular to a vertical line is horizontal

d A line perpendicular to a horizontal line is vertical

marble, not rolling

5B This wallpaper has a pattern of squares. How many sides of the four squares you can see in the picture are vertical, and how many are horizontal?

6C Each square is replaced by an equilateral (equal sided) triangle, and the same questions are asked. How many different answers are possible? Explain.

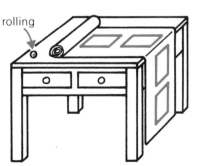

215

TURNING ABOUT A POINT

Time and trundling

===== Exercise 5 =====

1A What angle does the minute hand turn through between 4 o'clock and the following times?
  a 4.15   **b** 4.30   **c** 4.45

  d 4.05   **e** 4.10   **f** 5 o'clock

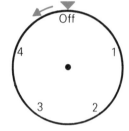

2A A cooker control has five positions, equally spaced round it.
  Through how many degrees does it turn in these changes of setting?
  a OFF to 1   **b** OFF to 2   **c** OFF to 3

  d OFF to 4   **e** 1 to 2   **f** 2 to OFF

3A

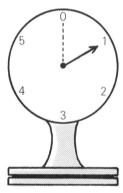

A weighing 'machine' scale is marked in kilograms.
How many degrees does the pointer turn through from zero to a weight of: **a** 1 kg **b** 3 kg **c** 5 kg?

4B Through what angle will the minute hand turn between 13 50 and the following times?
  a 14 00   **b** 14 05   **c** 14 20

  d 14 50   **e** 13 55   **f** 13 51

5C An angle may be bigger than one complete turn.
Through how many degrees will the hour hand have turned by the same time next day (Wednesday 8th)?
Through how many degrees will the minute hand have turned?
The second hand?

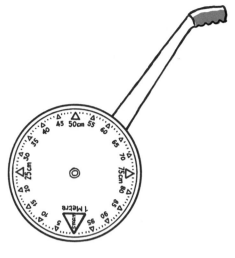

6C Through what angle will the trundle wheel turn in measuring these distances along the ground?

   a 25 cm   **b** 1 m   **c** 50 cm

   d 10 cm   **e** 5 cm   **f** 3 m

Money-go-round

=== *Exercise 6* ===

1A Which coin has turned least?

2A Which coin has turned most?

3A Place the coins in order from least turning to most turning.

4A Through how many degrees has the 50p coin turned?

5A Suppose the coins had all been turned **anticlockwise** instead of **clockwise** to reach their final positions.
Answer questions **1A–4A** again.

6A The coin has turned clockwise through how many degrees?

7A How much farther must it turn to get back to its starting position?

8A How many degrees will it then have turned through altogether?

9B Answer questions **6A**, **7A** and **8A** for the coin turning anticlockwise.

10C There are many more answers to question **7A**. Find some of them. Can you work out a rule for finding all of these answers?
Can you find the 'anticlockwise rule'?

1st position

2nd position

3rd position

4th position

11B Complete this table, writing down in degrees the smallest angle that the coin has turned through to get from the starting position to the finishing position. Some entries in the table are filled in for you. Check them first.

Finishing Positions

		1st Position	2nd Position	3rd Position	4th Position
Starting Positions	1st Position	0°	90°		
	2nd Position				180°
	3rd Position				
	4th Position			270°	

CLOCKWISE

12B Make up the same kind of table for anticlockwise turns.

Turn after turn

$0°$

Acute-angled turn
Between $0°$ and $90°$

Right-angled turn
$90°$; $\frac{1}{4}$-turn

Obtuse-angled turn
Between $90°$ and $180°$

Straight-angled turn
$180°$; $\frac{1}{2}$-turn

Reflex-angled turn
Between $180°$ and $360°$

Complete turn
$360°$

ANGLES AGAIN

=== *Exercise 7* ===

1A Give an example, in degrees, of an acute-angled turn, followed by another acute-angled turn, resulting in an obtuse-angled turn.

2A Give an example, in degrees, of an acute-angled turn, followed by an obtuse-angled turn, resulting in a straight-angled turn.

3A Write your results in the form of a table:

First turn	Second turn	Result	Example
acute	acute	obtuse	$23° + 80° = 103°$
acute	obtuse	straight	

4A Continue your table, using:
acute, right, obtuse, straight, reflex and complete turn.
Find as many different possibilities as you can, using these six different types of turn.

5C How many different combinations for the first and second turn do you think there are?
For each combination give examples of all possible results. Try to explain why the other results are not possible.

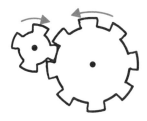

If one of these cogwheels turns, then it turns the other cogwheel. You will see this if you trace both cogwheels and cut them out in card.

This table shows the number of degrees turned by each wheel and the direction of turning.

Find x, y and z.

Can you find a rule for the number of degrees turned by the large wheel if the small wheel turns through $a°$?

Small wheel (4 cogs)	Large wheel (8 cogs)
360°	$x°$
180°	$y°$
$z°$	360°

Investigate these systems of cogwheels in the same way:

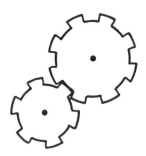

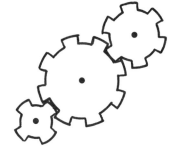

A 10p coin is rolled right round the edge of another 10p coin, which is kept fixed. What angle does it turn through?

Try this for other coins of the same size, and of different sizes.

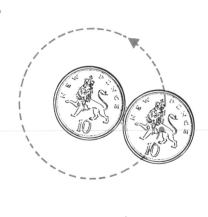

BEARINGS AND THE COMPASS

4 points

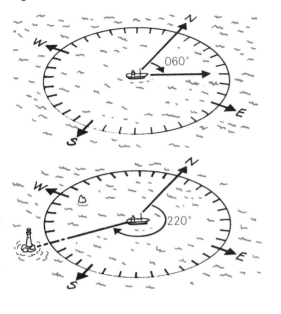

Bearings are measured from North in a clockwise direction.
They are always given in three figures.
The ship is sailing on a course bearing 060°, which is read 'zero six zero degrees'.

The lighthouse is on a bearing of 220° from the ship.

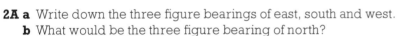
Exercise 8

1A Write down the three figure bearings of the following from O:

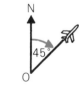
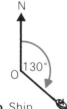
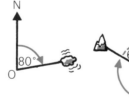
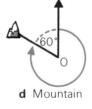

a Aircraft   **b** Ship   **c** Island   **d** Mountain

2A a Write down the three figure bearings of east, south and west.
   b What would be the three figure bearing of north?

3A Draw a circle with centre O. Mark the directions north, south, east and west, along with their three figure bearings.
Use a protractor to draw lines from O to show bearings of:
   a 030°   **b** 065°   **c** 100°   **d** 123°   **e** 200°   **f** 300°

4B In the ship and lighthouse diagram above, a rock bears 270° from the ship. What is the bearing of the ship from the rock?

5B The rock is due north of the lighthouse. What is the bearing of the lighthouse from the rock?

6C What is the bearing of the ship from the lighthouse?

7C Through what angle must the ship turn to follow a course due west?

8 points

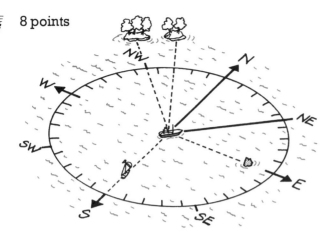

The ship is sailing north-east, and the yacht is sailing north.

Exercise 9

1A Write down the three figure bearings of:
  a north-east   **b** south-east   **c** south-west   **d** north-west

2A Give the three figure bearing of the ship's course.

3A What is the bearing of the rock from the ship?

4A a What is the bearing of the yacht from the ship?
  b What is the bearing of the ship from the yacht?

5A The yacht sends out a call for help. Through what angle would the ship have to alter course to sail towards the yacht?

6A What is the bearing of the 'Three tree island' from the ship?

7A What is the bearing of the 'Two tree island' from the ship?

8B What course should the ship follow if it is to pass between the two islands?

9C Estimate the bearing of the rock from the yacht, and of the yacht from the rock. How are these two bearings related to each other?

CHECK-UP ON **ANGLES AGAIN**

1A

Do you know these words?

acute | right | obtuse | straight | reflex | complete turn

horizontal | bearing | perpendicular | clockwise | vertical | anticlockwise

a *Copy and complete these labels:*

3° _____   135° _____   90° _____
360° _____   180° _____   230° _____

b *Copy and complete:*

N | W | E | S

(i) The string is _____. The horizon is _____. The string and the horizon are _____.

(ii) This man is facing north but is pointing on a _____ of 045°. To face this direction he would be quicker turning _____ not _____.

2B

Can you calculate angles?

Can you calculate bearings?

a 2 kg

a Through how many degrees will the pointer turn when this bag is weighed?

b N — Tom   N — Tim

b What is the bearing of Tim from Tom?

What is the bearing of Tom from Tim?

3C

Can you solve these problems?

a You are given a straight edge, compasses and a sheet of paper with torn edges. Draw a square on the paper.

b Arrange these swords to form 1 right angle and 2 obtuse angles.

18 | kg | 6 | 12

c When these four identical bags are placed on the scales the pointer turns through 240°. How much does one bag weigh?

4C

Investigate:

Tim

Tom

How can you work out Tom's bearing from Tim if you know Tim's bearing from Tom?

223

EARNINGS, EXPENSES AND SAVINGS

SAVING AND SPENDING

=== *Exercise 1A* ===

1 Each week Eric earns £4·75 delivering papers, and his Mum gives him £1·50 for doing odd jobs. On Saturdays he buys a football magazine for 85p and a record for £1·75. It costs £2·50 to go to the Match in the afternoon.
Copy and complete these tables for his earnings, expenses and savings.

Earnings	
Wages	
Odd jobs	

Expenses	
Magazine	
Record	
Football Match	

Savings	

2 Copy and complete Kim's earnings and expenses tables. How much can she save?

Earnings		Expenses	
Wages	£6·20	Make-up	£2·00
Tips	£1·75	Disco	£1·90
		Magazines	£1·50

3 Copy and complete Shalim's earnings and expenses tables.

Earnings		Expenses	
Wages	£8·00	Football Match	£1·90
Grass cutting	£1·50	Disco	£1·50
		Bus fares	£0·70

Calculate his savings.

4 One week Sally earned £7·38 by working part-time in a shop. Her aunt gave her £2 for baby-sitting. She bought a jumper for £5·17, a record for £1·67 and a poster for 95p. She put the rest in her building society account. Make out earnings and expenses tables for Sally. How much did she put into her account?

5 Gavin has just left school and is on a Young Adults Programme. He works in a double-glazing factory for a wage of £42·50 a week, and is given £3·75 towards his travelling costs. Each week he gives £14 to his parents, spends £7·50 on lunches and £4·80 on bus fares. Make up earnings and expenses tables, and find how much he can save.

6 John Hudson takes home £77·20 per week from his job as a car mechanic. He gives his parents £22·50 and spends £1·85 per day on lunch in the canteen from Monday to Friday. His car costs, on average, £13·50 a week to run. How much of his weekly wage is left for spending or saving?

7 Imagine you have left school and have got a job. You may be living at home, or perhaps sharing a flat. Make up your own earnings, expenses and savings budget, as you think it would be.

Exercise 1B

1 Heather Templeton works in a boutique five and a half days a week and has a take-home pay of £66·75. Her expenses are £21 a week to her mother, a daily train ticket for £1·24, morning coffee at 35p and £7·50 a week for lunches. What are Heather's total expenses for the week? How much money has she left each week to spend or save?

2 Alison's sister Chris has just started University. She gets a grant of £1650 to cover the 30 weeks of the University terms. She will need £300 for books and £250 for other expenses. How much does this leave?

During the weeks she attends University, she expects to spend £8 a week on lunches and £8·64 a week on train fares.

How much of her grant is left now?

3 Mr Wilson works in a bank, and receives a monthly cheque for £621·68. Mrs Wilson is a part-time teacher, and has a monthly cheque for £185·22. Each month they have the following bills to pay:

House mortgage	£236·50	Electricity	£ 28·75
House insurance	£ 17·78	Gas	£ 43·50
Life insurance	£ 52·34	Rates	£105·37

Their car costs £1050 a year to run, and they set aside an equal part of this monthly to cover the cost. How much is left each month to buy food, clothes, etc?

4 Hill Dean School held a concert to raise money for school funds. 284 tickets at £1·50 each and 167 programmes at 25p each were sold. The costs of the show were £38·75 for scenery, £34 for costumes, £12·35 for programmes and £5 for producing the tickets. Make income and expense tables and find how much money was raised.

DINING OUT

SAVING AND SPENDING

TASTY BITE RESTAURANT

MENU

STARTERS
Fruit juice	50p
Melon boat	75p
Soup	55p

MAIN MEAL
SALADS
Slimmer's	£2-75
Gammon	£3-15
Turkey	£3-75

Roast beef and Yorkshire pudding	£4-50
Shepherd's pie	£2-65
Chicken curry	£3-40
Smoked trout	£4-15
Haddock Orly	£3-50

FRENCH FRIES and CHOICE of VEGETABLE INCLUDED

SWEETS
Peach melba	80p
Apple pie and custard	75p
Assorted ices	65p
Chocolate gateau	£1-10

DRINKS
Coffee	60p	Milk	25p
Tea	40p	Minerals	35p

TASTY BITE

BILL

Item	£ p
1 soup	55
2 gammon salad	6 30
1 peach melba	80
2 coffees	1 20
	8 85

PAID WITH THANKS

===== Exercise 2A =====

1 Miss Stewart had lunch at the Tasty Bite Restaurant. She chose: melon boat, haddock orly and tea. Make out her bill like the one shown above.

2 Mr Big and Mr Boss, two businessmen, are having lunch at the Tasty Bite. Mr Big has soup, roast beef, gateau and coffee. Mr Boss has tomato juice, smoked trout, peach melba and tea. Make out a separate bill for each of them.

3 Ann and Eleanor are watching their weight, so each chooses fruit juice and slimmer's salad. Ann has an ice, but Eleanor weakens and has peach melba. Make out a bill for each of them.

4 Make a list of the prices in the school canteen. Plan your own meals for a week, and calculate the cost. Have you chosen a healthy diet?

===== Exercise 2B =====

1 John, Tina and their mum and dad are eating out at the Tasty Bite. It is Tina's birthday. She has juice, curry, ice-cream and lemonade. John has juice, turkey salad, peach melba and milk. Mum and dad both decide to have soup, shepherd's pie, apple pie and tea. Make out a bill for the whole family.

2 If you choose one starter, one main course, one sweet and one drink, what is the cost of the dearest possible meal? What is the cost of the cheapest meal?

3 Ronald is buying lunch for his fiancée, Fiona. They are both very hungry, but Ronald has only £8 to spend. Decide what they can have to eat, and make out their bill.

AT THE SUPERMARKET

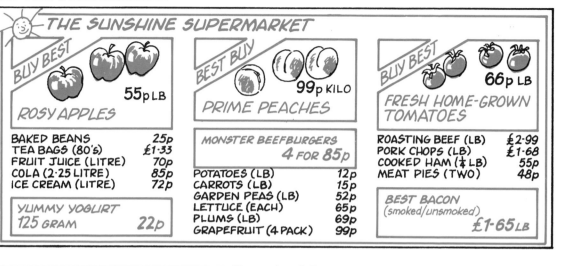

THE SUNSHINE SUPERMARKET

ROSY APPLES   55p LB

PRIME PEACHES   99p KILO

FRESH HOME-GROWN TOMATOES   66p LB

BAKED BEANS	25p
TEA BAGS (80's)	£1·33
FRUIT JUICE (LITRE)	70p
COLA (2·25 LITRE)	85p
ICE CREAM (LITRE)	72p

YUMMY YOGURT
125 GRAM   22p

MONSTER BEEFBURGERS
4 FOR 85p

POTATOES (LB)	12p
CARROTS (LB)	15p
GARDEN PEAS (LB)	52p
LETTUCE (EACH)	65p
PLUMS (LB)	69p
GRAPEFRUIT (4 PACK)	99p

ROASTING BEEF (LB)	£2·99
PORK CHOPS (LB)	£1·68
COOKED HAM (¼ LB)	55p
MEAT PIES (TWO)	48p

BEST BACON
(smoked/unsmoked)   £1·65 LB

Exercise 3A

Use the supermarket advertisement to calculate the cost of these shopping lists. Write each one in the form of a bill—item, then cost, then total.

1
1 lb apples
1 kilo peaches
1 fruit juice
1 lettuce
1 lb roast beef

2
1 lb tomatoes
4 beefburgers
1 pkt teabags
1 yogurt
1 lb peas

3
3 lb plums
1 litre ice cream
2 bottles cola
4 grapefruit
2 lb bacon
2 lettuce

4
3 lb apples
2 yogurts
½ lb peas
2 lb pork chops
2 kilo peaches
6 meat pies

5 Make up and cost a shopping list of your own choice for a meal for 4 people.

Exercise 3B

Here are the prices of some items on sale at a supermarket last year.

a
125 ml	45p
250 ml	79p

b
500 ml	44p
1 litre	87p

c
200 ml	19p
1 litre	59p

d
pkt 150 g	65p
box 146 g	69p

e
50 g	80p
100 g	138p
200 g	260p

f
340 g	52p
680 g	94p
2 kg	218p

g
6 for	62p
36 for	252p

h
10 pack	68p
5 pack	33p

1 Look at the quantities and the prices. Say whether or not you think that the larger pack is better value in each case.

2 Check your answer by calculating the cost of 1 ml or 1 g for the packs in **a**, **b**, **c** and **f**.

3 Kitchen foil is sold in three different packs. How would you decide which to buy?

450 mm by 9 m   89p
450 mm by 4·5 m   53p
300 mm by 9 m   67p

=========== *Exercise 3C* ===========

A computer-controlled cash-register in a shop gives change to customers, using the least possible number of coins.

The flowchart shows you how to do this.

'A' stands for the amount of change to be given.

'C' stands for the coin being considered.

$\boxed{A = A - C}$ replaces the existing value of A

by a new value, $A - C$, and calls this new value A again.

If $C = 20$, then $\boxed{\text{Write 'Cp coin'}}$ means that you

write down '20p coin'.

1 Use the flowchart carefully to process the 73p change.

2 In the first process box replace 73 by 16, and follow through the method of getting 16p change.

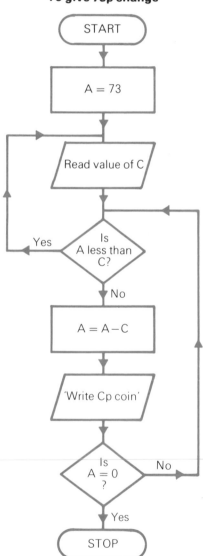

To give 73p change

START

A = 73

Read value of C

Is A less than C? —Yes

No

A = A − C

'Write Cp coin'

Is A = 0 ? —No

Yes

STOP

Data: 50, 20, 10, 5, 2, 1

MILK DELIVERY

Mike the milkman delivers milk, yogurt and cream daily to his customers, and each Friday he collects the money. He uses a ready reckoner like this to help him to calculate each bill.

		Price	
Number	Milk (In Pints)	Yogurt (In Cartons)	Cream (In Cartons)
1	26p	24p	66p
2	52p	48p	£1·32p
3	78p		
4			
5			
6			

================= *Exercise 4A* =================

1 Copy and complete Mike's ready reckoner.

2 Use the ready reckoner to calculate the cost of the following orders:

	Milk (pints)	**Yogurt** (cartons)	**Cream** (cartons)		**Milk** (pints)	**Yogurt** (cartons)	**Cream** (cartons)
a	5	—	—	**d**	3	2	1
b	3	1	—	**e**	10	—	—
c	4	—	2	**f**	8	6	2

3 Calculate the weekly milk bill for your own home.

================= *Exercise 4B* =================

Calculate the cost of the following orders by means of the ready reckoner:

1 6 pints of milk, 5 cartons of yogurt and 4 of cream.

2 9 pints of milk and 6 cartons of cream.

3 5 pints of milk, 8 cartons of yogurt and 2 of cream.

4 Mike could calculate the cost of 10 pints by adding the costs of 5 and 5 pints, or 6 and 4 pints. List all the prices he could add together for:
   a 9 pints   **b** 8 pints   **c** 7 pints.

5 Ready reckoners were very useful before electronic calculators were invented. Use your calculator to work out some of the questions above. What do you think about the two methods?

ON HOLIDAY

SAVING AND SPENDING

=========== *Exercise 5* ===========

1 The Seaview Camping Site has the following charges:

Adults (16 and over)......: £1·10 each per night
Children (15 and under): 50p each per night

Also

Tent ....: £2·10 per night for hire
Caravan ....: £3·50 per night for hire

---

a What is the charge for: (i) a 14-year-old staying one night *and* hiring a tent
 (ii) two adults staying one night *and* hiring a caravan?
b Sean, Gareth and David are on a camping holiday. Sean and Gareth are 18, and David is 15. They hire a tent for one night. What will they be charged?
 They decide to stay 3 more nights. How much will these three nights cost them?
c Mr and Mrs Paterson and their 8-year-old son spend 5 nights in a hired caravan at Seaview. How much will they pay?

2 The Fernlea Boarding House charges as follows:

Bed and Breakfast Adult—£12·30   Under 14—$\frac{1}{2}$ price.

Evening meal   Adult—£5·40   Under 14—$\frac{1}{2}$ price.

a How much does it cost Mr and Mrs Graham to stay for 5 nights and have a meal each evening?
b Mr and Mrs Jones arrive at Fernlea with their son Tom (15 years old) and their daughter Alice (12 years old). They want to know what the charge will be to stay for 4 nights and have 3 evening meals. What is the cost?

3 Try to work out the cost of a week's holiday for a family of your own choice staying in a hired caravan at Seaview, and buying their own food.
Compare it with the cost at Fernlea, having evening meals, and buying lunches daily.

4 A geography teacher takes 8 pupils in the school mini-bus on a field study trip to the Scottish Highlands for 5 days. During the trip they travel 600 miles in the mini-bus which uses 1 litre of fuel every 5 miles. 1 litre of fuel costs 52 pence. They rent a cottage for £58 and cook their own meals.
If the cost of the fuel and the cottage is divided equally among the whole party; and each pays £25 for food, how much does the trip cost each person?

This table comes from a holiday brochure. It gives the cost of a package holiday to the 'Olé' Holiday Camp on the Costa del Sol in Spain. The prices for each person in the table include air fare, accommodation and all meals.

	May	June	July	Aug	Sept	Oct
7 nights—**Adult**	£219	£231	£272	£265	£238	£205
7 nights—**Child**	145	158	184	178	160	133
14 nights—**Adult**	269	305	343	338	295	265
14 nights—**Child**	174	216	238	224	197	168

Adult: 15 and over   **Child:** Under 15.

1 From the table, write down the cost of a holiday for:
   a 1 adult for 7 nights in June   **b** 1 child for 7 nights in August
   c 1 adult for 14 nights in October   **d** 1 child for 14 nights in July.

2 a What is the cheapest 7-night charge for an adult? In what month?
   b What is the dearest 7-night charge for an adult? In what month?
   c Why are there different charges?

3 Mr and Mrs Foster book places at the camp for 7 nights in October. How much will they be charged?

4 The Gilbert family want to go to the Holiday Camp during the first fortnight in June. Mr and Mrs Gilbert have two children, one aged 9 and the other aged 7. How much will they have to pay?

5 Mr and Mrs Johnstone have three children, aged 15, 13 and 10.
   If they all went on a package holiday to the camp from the 5th to the 12th of July, how much would they be charged?

6 The Olé Camp has special reduced prices for large parties, to try to attract more people. For each group of 12 paying adults, one extra person is allowed free.
   Twenty-seven members of the Sunnyside Old Folks Club decide to go to the camp for the last week in May. How much will the club be charged? How much should each member of the club pay?

7 a How much would it cost your family to spend 7 nights at the camp in August?
   b Estimate how much you would need to save to have a reasonable amount of spending-money for your holiday.

FLY to Paris from your LOCAL AIRPORT for a SUMMER/WINTER BREAK

SAVING AND SPENDING

	October 31—March 31			April 1—October 30		
	3 days 2 nights	4 days 3 nights	8 days 7 nights	3 days 2 nights	4 days 3 nights	8 days 7 nights
Cardiff & Bristol	—	—	—	£159	£168	£208
Southampton	£122	£129	£165	132	143	181
Stanstead	113	120	156	126	135	174
Birmingham	143	150	196	153	163	202
East Midlands	143	150	196	153	163	202
Manchester	154	161	200	164	175	220
Leeds/Bradford	192	199	241	195	215	253
Newcastle	195	204	235	205	218	260
Glasgow	186	195	230	189	211	248
Edinburgh	198	210	244	206	220	261
Aberdeen	201	214	248	210	225	266

Prices quoted are for adults: Children under 12—$\frac{1}{5}$ reduction.

1 How much would the holiday cost Sam and Susan, flying from Glasgow for a 4 day package at the end of August?

2 Mr and Mrs Simpson and their 14-year-old son are flying from Birmingham for an 8 day holiday in September. How much will they pay?

3 Mr and Mrs Macdonald, their son Ian aged 12 and daughter Yvonne aged 10, plan to go to Paris for 3 days in December. If they fly from Manchester, what would the package holiday cost?

4 Mr and Mrs Peters and their 5-year-old twin daughters are flying from Bradford for an 8 day Easter break in Paris in March. How much will this cost them?

5 Choose one of the package holidays in the table, and find the cost for your own family.

Collect some holiday brochures. Working on your own, or with others in the class, decide where and when you would like to go, if you had the chance.
Calculate the cost of the holiday as accurately as possible.

CHECK-UP ON **SAVING AND SPENDING**

1A, B Earnings, expenses and savings

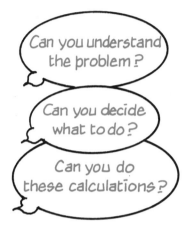

a A golfer won £10 350 in a competition.
His expenses were: Hotel £212·50
Travel £83·25
Caddy—one tenth
of prize money.
He bought a new set of golf clubs for £438·75.
How much of his prize money was left?

b Find the total cost of:
1 kilo peaches at 95p a kilo
2 pkts cereal at 88p a pkt
4 yogurt at 23p each
2 lb bacon at £1·75 lb
3 bars of Choco ice at 29p each.

2A, B Ready reckoners

Here are the prices of spring bulbs from the Petal Palace nursery:

Bulb \ Number	10	25	50	100
Daffodil	£1·55	£3·80	£7·15	£13·80
Tulip	2·45	6·00	11·00	20·50
Crocus	1·25	3·00	5·75	10·15
Snowdrop	0·65	1·50	3·45	6·25

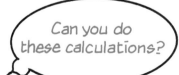

Calculate the cost of:

a 50 daffodils and 10 tulips

b 100 crocus and 25 daffodils

c 25 tulips and 100 snowdrops

d 50 snowdrops and 75 crocus

e 100 daffodils, 150 crocus and 75 tulips.

233

FORMING FORMULAE

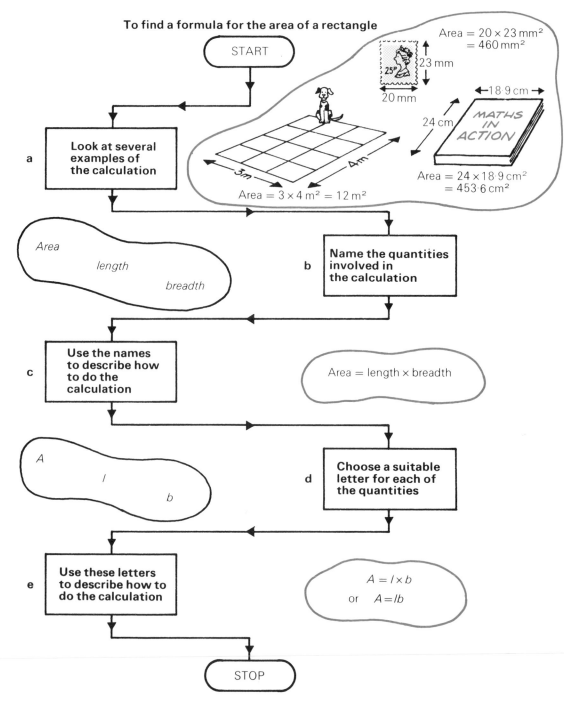

To find a formula for the area of a rectangle

START

a Look at several examples of the calculation

Area = 20 × 23 mm²
= 460 mm²

23 mm

20 mm

←18·9 cm→

24 cm

MATHS
IN
ACTION

Area = 24 × 18·9 cm²
= 453·6 cm²

3m 4m

Area = 3 × 4 m² = 12 m²

Area

length

breadth

b Name the quantities involved in the calculation

c Use the names to describe how to do the calculation

Area = length × breadth

A

l

b

d Choose a suitable letter for each of the quantities

e Use these letters to describe how to do the calculation

A = l × b

or   A = lb

STOP

We have found the formula $A = lb$ for the area of a rectangle. This can now be used to calculate the area of any rectangle.

234

1A a Calculate the **change** you would receive from the **money** and the **price** in each picture—copy and complete a table like this:

Money	£10	£5	
Price	£7		
Change			

b Name the three quantities in each example.
c How did you calculate the change?
d Choose letters for the three quantities. (*C, M, P.*)
e Write down a formula for calculating the change; '*C* = ...'

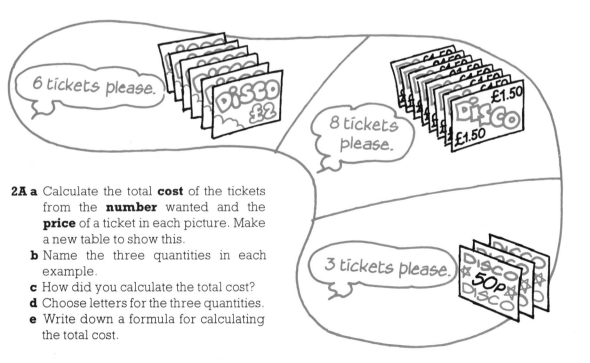

2A a Calculate the total **cost** of the tickets from the **number** wanted and the **price** of a ticket in each picture. Make a new table to show this.
b Name the three quantities in each example.
c How did you calculate the total cost?
d Choose letters for the three quantities.
e Write down a formula for calculating the total cost.

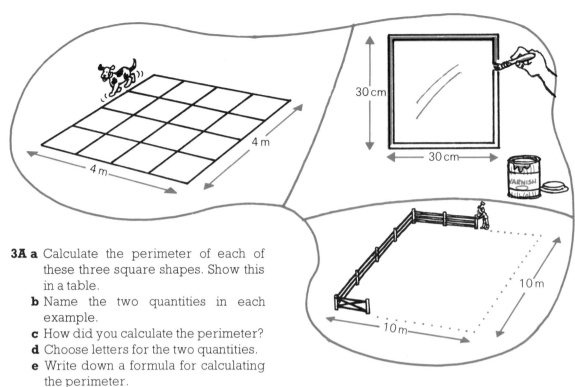

3A a Calculate the perimeter of each of these three square shapes. Show this in a table.
b Name the two quantities in each example.
c How did you calculate the perimeter?
d Choose letters for the two quantities.
e Write down a formula for calculating the perimeter.

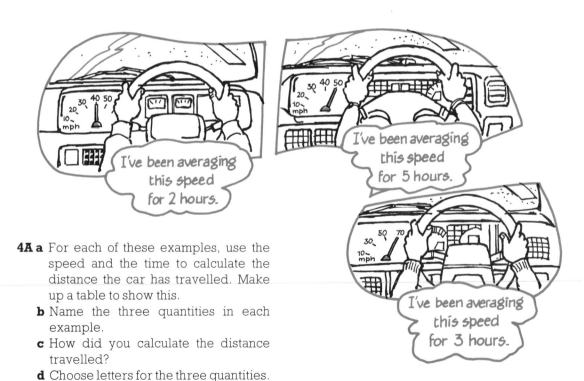

4A a For each of these examples, use the speed and the time to calculate the distance the car has travelled. Make up a table to show this.
b Name the three quantities in each example.
c How did you calculate the distance travelled?
d Choose letters for the three quantities.
e Write down a formula for calculating the distance travelled.

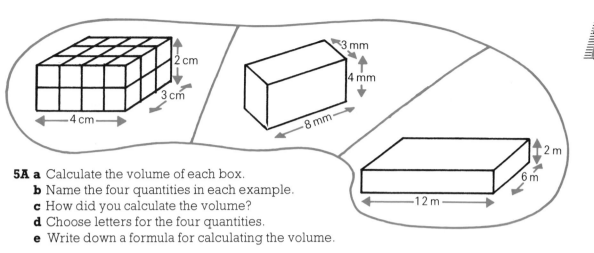

5A a Calculate the volume of each box.
 b Name the four quantities in each example.
 c How did you calculate the volume?
 d Choose letters for the four quantities.
 e Write down a formula for calculating the volume.

6B a Calculate the number of 20 minute programmes that can be recorded on each of these
   video tapes.
 b Name the three quantities you used.
 c How did you calculate the number of programmes?
 d Choose letters for the three quantities.
 e Write down a formula for calculating the number of 20 minute programmes.
 f Write down a formula for calculating the number of x minute programmes.

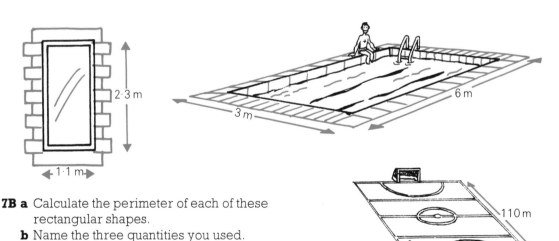

7B a Calculate the perimeter of each of these
   rectangular shapes.
 b Name the three quantities you used.
 c Explain how you calculated the perimeter.
 d Choose letters for the three quantities.
 e Write down a formula for calculating the perimeter.

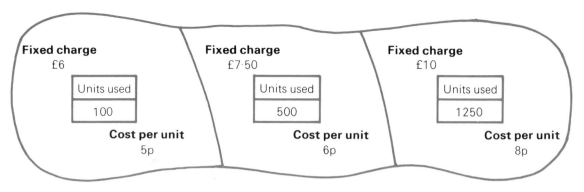

Fixed charge £6	Fixed charge £7·50	Fixed charge £10
Units used 100	Units used 500	Units used 1250
Cost per unit 5p	Cost per unit 6p	Cost per unit 8p

8C a Calculate the electricity bill in each example.
 b Name the four quantities involved.
 c How did you calculate the bill?
 d Choose letters for the four quantities.
 e Write down a formula for calculating the electricity bill.

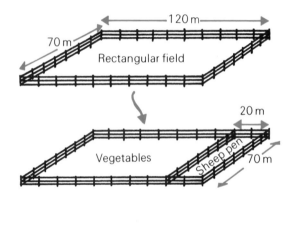

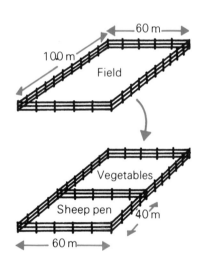

9C a Calculate the area left for vegetables after the sheep pen has been fenced off in each of these fields.
 b Name the four quantities you used.
 c Explain your calculation.
 d Find a formula for calculating the area of the vegetable patch.
 e Find the formula if the areas of the vegetable plot and the sheep pen are equal.

10C Find a formula for calculating the total length of fence needed to enclose the vegetables and sheep in the examples of question **9C**.

FURTHER FORMULAE

Shortcuts

$a \times b$ is written ab.

$2 \times n = 2n$.

$c \times c$ is written c^2. This is read 'c squared'.

Exercise 2

In each question: **a** Copy and complete the calculations.

b Use the letters in brackets to make a formula.

Given quantity	Quantity to be found	Calculations
1A Number of hours (H).	Number of minutes (M) in H hours.	Number of minutes in 1 hour = 60 ,, ,, ,, ,, 5 hours = 5 × 60 = ,, ,, ,, ,, 10 hours = = ,, ,, ,, ,, H hours = $H \times 60 = 60H$ The formula is $M = 60H$
2A Number of £s (P).	Number of pence (p) in £P.	Number of pence in £1 = ,, ,, ,, ,, £9 = ,, ,, ,, £15 = ,, ,, ,, ,, £P = The formula is $p =$ _____
3A Number of centimetres (C).	Number of millimetres (M) in C cm.	Number of mm in 1 cm = 10 ,, ,, ,, ,, 7 cm = ,, ,, ,, ,, 12 cm = ,, ,, ,, ,, C cm = The formula is _____
4A Number of years (Y).	Number of days (D) in Y years.	Number of days in 1 year = ,, ,, ,, ,, 2 years = ,, ,, ,, ,, 3 years = ,, ,, ,, ,, Y years = The formula is _____
5A Number of kilometres (K).	Number of metres (M) in K km.	Number of metres in 1 km = 1000 ,, ,, ,, ,, 3 km = ,, ,, ,, ,, 6 km = ,, ,, ,, ,, K km = The formula is _____

FORMULAE

Given quantity	Quantity to be found	Calculations

6A Length (l cm) and breadth (b cm) of the rectangle.

Area of the rectangle (A cm²).

Area
$= 3 \times 2$ cm²
$= 6$ cm²

Area
$= \ldots\ldots$
$= \ldots\ldots$

Area
$= \ldots\ldots$

The formula is _____

7A Length of a side of the square (s mm).

Area of the square (A mm²).

Area
$= 2 \times 2$ mm²
$= \ldots\ldots$

Area
$= \ldots\ldots$
$= \ldots\ldots$

Area
$= \ldots\ldots$

The formula is _____

8A Length of a side of the cube (c cm).

Volume of the cube (V cm³).

Volume
$= 2 \times 2 \times 2$ cm³
$= \ldots\ldots$

Volume
$=$
$= \ldots\ldots$

Volume
$=$

The formula is _____

9B Length of a side of the cube (c cm).

Total surface area of the cube (S cm²).

Surface area
$= 6 \times 4$ cm²
$= \ldots\ldots$

Surface area
$=$
$= \ldots\ldots$

Surface area
$=$

The formula is _____

10B Number of seconds (s).

Number of minutes (m).

Number of minutes in 60 seconds $=$
,, ,, ,, ,, 120 ,, $=$
,, ,, ,, ,, 300 ,, $=$
,, ,, ,, ,, s ,, $=$

The formula is _____

11B

Time taken, in minutes, to read the book (t).

Number of pages	Pages per minute	Time taken in minutes
100	2	$\frac{100}{2} = 50$
600	3	
256	4	
p	m	

The formula is _____

Given quantity	Quantity to be found	Calculations

12B

The record score (R)
Your score so far (S)

Points needed to
equal the record (P).

Record score	Your score	Points needed
8200	7200	
10 000	9500	
11 000	2222	
R	......	

The formula is _____

13B

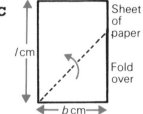

The number of miles
per litre of petrol (m)

The number of litres of
petrol (n) the tank can
hold when full

The number of miles
the car can travel
on a full tank (f).

Number of litres tank can hold	Number of miles per litre	Total number of miles
45	6	
40	11	
64	12	
......	......	

The formula is _____

14C

Sheet of paper

Fold over

Cut, and open out folded part

Remainder

Square

The length of the
paper (l cm).
The breadth of the
paper (b cm).

The area of
the remainder
(A cm²).

Length (cm)	Breadth (cm)	Area (cm²)	Area of square (cm²)	Area of remainder (cm²)
8	6	48	36	
10	9			
12	12			
......	......			

15C

The number of hens (h).
The average number of
eggs each hen lays per
day (e).
The number of days (d).
The selling price of each
egg (p) in pence.

The amount of
money made from
selling the eggs (M)
in £.

a 10 hens
1 egg per day
5 days
7p per egg

b 100 hens
0·9 eggs per day
10 days
6·2p per egg

FORMULAE

FINAL FORMULAE

Making and using a formula

=== *Exercise 3* ===

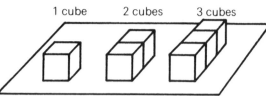

1A Cubes are placed on a table, like this, and then they are painted. The base of each cube is left unpainted.
Copy and complete these tables:

Number of cubes (c)	1	2	3	4	5	6
Number of square faces to be painted (s)	5	8	11			

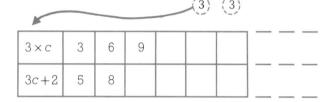

$3 \times c$	3	6	9			
$3c+2$	5	8				

The formula for calculating the number of squares to be painted is $\boxed{s = 3c+2.}$

2A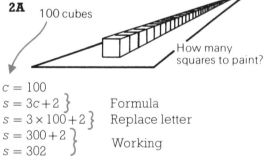
100 cubes
How many squares to paint?

$c = 100$
$\left. \begin{array}{l} s = 3c+2 \end{array} \right\}$ Formula
$\left. \begin{array}{l} s = 3 \times 100 + 2 \end{array} \right\}$ Replace letter
$\left. \begin{array}{l} s = 300 + 2 \\ s = 302 \end{array} \right\}$ Working
There are 302 squares to paint.

In the same way use the formula $s = 3c+2$ to calculate the number of squares (s) to be painted for:
a 10 cubes ($c = 10$)   **b** 20 cubes
c 11 cubes   **d** 18 cubes

3A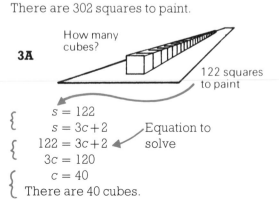
How many cubes?
122 squares to paint

$\left\{ \begin{array}{l} s = 122 \\ \quad s = 3c+2 \\ 122 = 3c+2 \\ \quad 3c = 120 \\ \quad\quad c = 40 \end{array} \right.$ Equation to solve
There are 40 cubes.

In the same way, use the formula $s = 3c+2$ to calculate the number of cubes (c) with:
a 23 painted squares ($s = 23$)
b 32 painted squares
c 35 painted squares.

4A Cubes are placed on a table, like this, and then they are painted.

1 cube   2 cubes   3 cubes

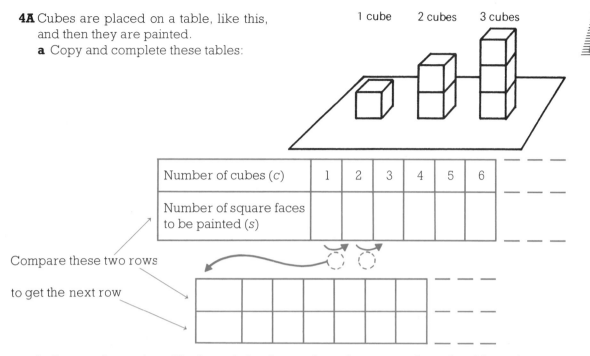

a Copy and complete these tables:

FORMULAE

Number of cubes (c)	1	2	3	4	5	6
Number of square faces to be painted (s)						

Compare these two rows

to get the next row

b Copy and complete: The formula for the number of squares to be painted is $s = 4c + \ldots\ldots$

5A Use your formula in question **4A** for these:

a
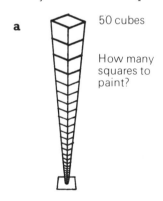
50 cubes

How many squares to paint?

b How many squares have to be painted for:

(i) 8 cubes   (ii) 12 cubes   (iii) 20 cubes?

6A Use your formula in question **4A** for these:

a

401 squares to paint

How many cubes?

b Calculate the number of cubes with:

(i) 37 painted squares

(ii) 29 painted squares

(iii) 1001 painted squares.

7A Cubes are placed on a table, and against a wall, like this. Then they are painted.
a Copy and complete these tables:

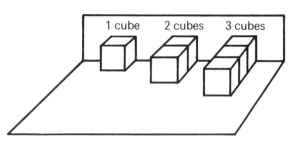

| 1 cube | 2 cubes | 3 cubes |

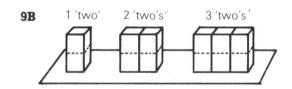

Number of cubes (c)	1	2	3			
Number of squares (s)						

b Write down a formula for s in terms of c.

8A Use your formula in question **7A** for these:

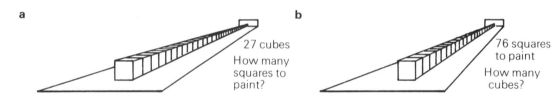

a

27 cubes
How many squares to paint?

b

76 squares to paint
How many cubes?

9B

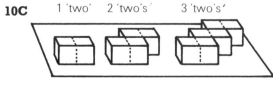

1 'two'   2 'two's'   3 'two's'

a By making tables as you have done in the previous questions, find a formula connecting the number of squares (s) and the number of 'twos' (t).

b For 24 'twos' how many squares are there to paint?
c How many 'twos' would there be if there are 64 squares to paint?

10C

1 'two'   2 'two's'   3 'two's'

a Find a formula connecting the number of squares to be painted (s) and the number of 'twos' (t).
b How many squares are there to paint if 1000 'twos' are arranged like this?
c Is it possible to arrange 'twos' in this way so that there are 104 squares to be painted? Explain your answer.

1A

Do you know these words?

plus

quantities

formula   squared

Copy and complete:

$R = S + V^2$ is a _____ with three different _____. The formula says: 'R equals S _____ V _____.'

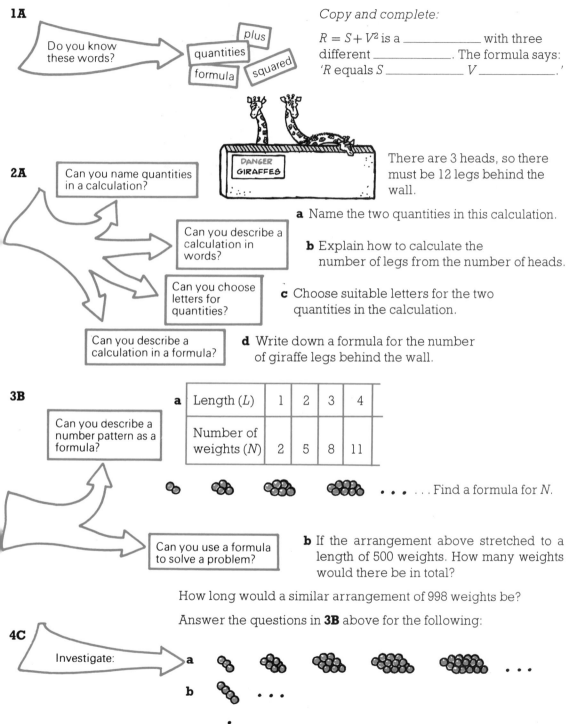

2A

Can you name quantities in a calculation?

Can you describe a calculation in words?

Can you choose letters for quantities?

Can you describe a calculation in a formula?

DANGER GIRAFFES

There are 3 heads, so there must be 12 legs behind the wall.

a Name the two quantities in this calculation.

b Explain how to calculate the number of legs from the number of heads.

c Choose suitable letters for the two quantities in the calculation.

d Write down a formula for the number of giraffe legs behind the wall.

3B

Can you describe a number pattern as a formula?

Can you use a formula to solve a problem?

a	Length (L)	1	2	3	4
	Number of weights (N)	2	5	8	11

. . . . Find a formula for N.

b If the arrangement above stretched to a length of 500 weights. How many weights would there be in total?

How long would a similar arrangement of 998 weights be?

Answer the questions in **3B** above for the following:

4C

Investigate:

a   . . .

b   . . .

:

INTERPRETATION

Meet the people

Ian   Tom

Martin   Alistair

Ian is older and taller
than Tom.
We can show this on a graph:

Martin is taller than Alistair.
Alistair is older than Martin.
We can show this on a graph:

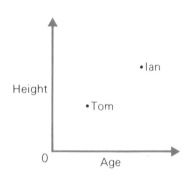

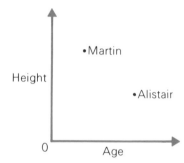

GRAPHS AND RELATIONS

═══════════ *Exercise 1* ═══════════

1A Look at the picture and then at the graph.

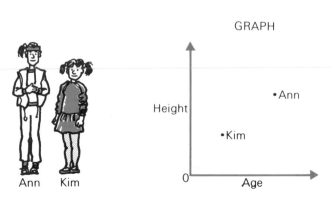

GRAPH

a Which girl is taller?
b Which girl is younger?

2A

Murray John

GRAPH

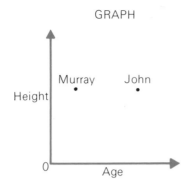

Height — Murray • ; John •

0 — Age

a Who is the older man?

b What can you say about the height of the men?

c 'I was born before you, Murray' said John. Is he correct?

3A

Peter Douglas

GRAPH

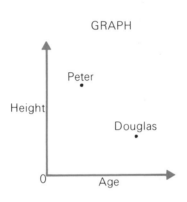

Height — Peter •

Douglas •

0 — Age

a On the graph why is the point for Peter above the point for Douglas?

b Why is the point for Peter to the left of the point for Douglas?

4B

Bob Grant Finlay

GRAPH

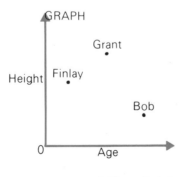

Grant

Height — Finlay

Bob

0 — Age

a Who is the shortest?

b Who is the youngest?

c List these three people in order from youngest to oldest.

d Now list them in order from shortest to tallest.

e Why is the point for Bob to the right of the points for Grant and Finlay on the graph?

f Why is the point for Finlay below the point for Grant, but above the point for Bob?

GRAPHS AND RELATIONS

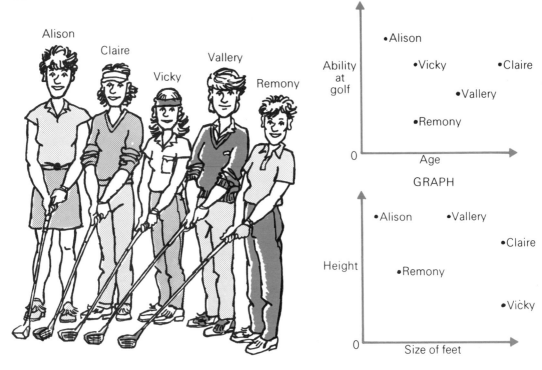

GRAPH

Ability at golf

• Alison
• Vicky   • Claire
          • Vallery
• Remony

0   Age

GRAPH

Height

• Alison   • Vallery
                    • Claire
• Remony

          • Vicky

0   Size of feet

Copy these tables and then use the graphs to complete them.

Shoe size	Name
3	
4	
$4\frac{1}{2}$	
6	
6	

Age (years)	Name
19	
20	
20	
21	
23	

Height (cm)	Name
159	
168	
175	
186	
186	

Handicap	Name
24	
20	
18	
18	
10	

(The better the golfer the lower the handicap.)

Name the trees

=== *Exercise 2* ===

1A

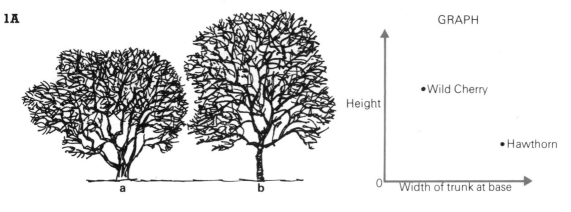

GRAPH

Height

• Wild Cherry

• Hawthorn

0   Width of trunk at base

Which is the wild cherry? Use the graph to help you to decide.

2A

GRAPH

• Lime

• Larch

Height

• Crab apple

0   Width of trunk at base

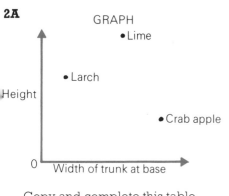

Copy and complete this table.

Tree	a	b	c
Name			

3A

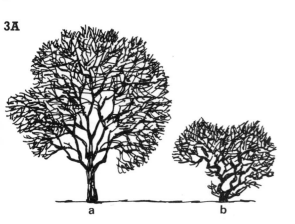

GRAPH

Height

• Hornbeam

• Tamarisk

0   Age

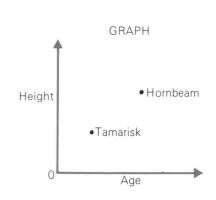

Which tree is older?
Is the younger
tree on the left?

249

GRAPHS AND RELATIONS

4A

GRAPH

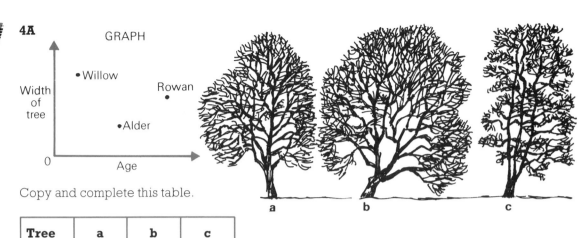

Width
of
tree

• Willow

Rowan
•

•Alder

0   Age

a   b   c

Copy and complete this table.

Tree	a	b	c
Name			

Is the widest tree also the oldest tree?
Are the oldest and youngest trees next to each other?

5B

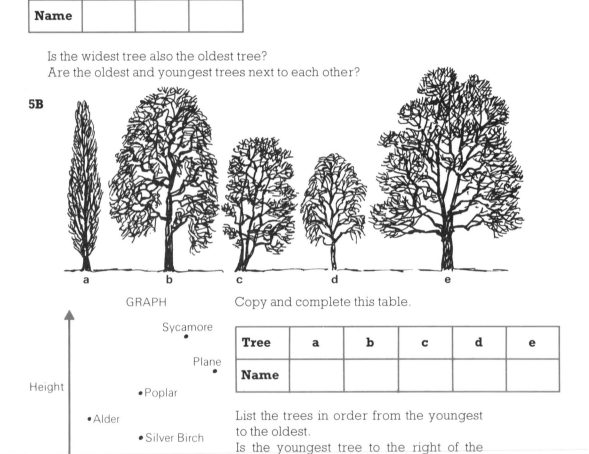

a   b   c   d   e

GRAPH

Sycamore
•

Plane
•

Height

•Poplar

•Alder

• Silver Birch

0   Age

Copy and complete this table.

Tree	a	b	c	d	e
Name					

List the trees in order from the youngest
to the oldest.
Is the youngest tree to the right of the
oldest tree?

Which of the statements below are true and which are false?

The oldest and youngest
trees are next to each other.

There is one tree
between the highest
and the youngest tree.

The oldest tree is
also the highest.

LABELLING

Compare the coins

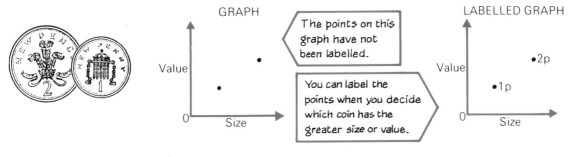

GRAPH

Value

0   Size

> The points on this graph have not been labelled.

> You can label the points when you decide which coin has the greater size or value.

LABELLED GRAPH

Value

•2p

•1p

0   Size

Exercise 3

Copy and label the graphs in questions **1A–8A**

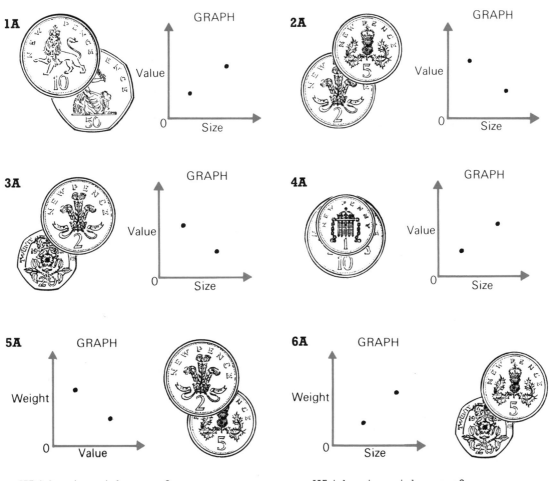

1A

GRAPH

Value

0   Size

2A

GRAPH

Value

0   Size

3A

GRAPH

Value

0   Size

4A

GRAPH

Value

0   Size

5A   GRAPH

Weight

0   Value

Which coin weighs more?

6A   GRAPH

Weight

0   Size

Which coin weighs more?

7A

GRAPH

Value

0   Weight

Which coin weighs more?

8A

GRAPH

Size

0   Weight

Which coin weighs more?

9B

GRAPH
Coins used in Britain

Weight

0   Value

Label these points.
Then list the coins
in order from
lightest to heaviest.

10C

GRAPH
Coins used in New Zealand

Weight

0   Value

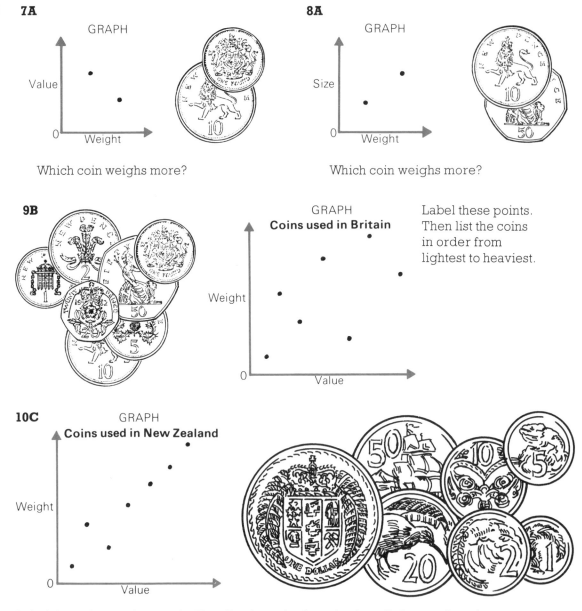

Label the points on the graph. Then list the coins in order from lightest to heaviest.
New Zealanders think their system of coins is better than the British system. Can you see
why? Some Britons think their system is more practical. Do you agree? Why?

From your work on questions **9B** and **10C** what advice would you give to the
Royal Mint on choosing a system of coins? Remember that Britain has a decimal
system of counting, and keep in mind the needs of shoppers and shopkeepers, as
well as tourists, blind people and others.

PLOTTING

Develop the plot

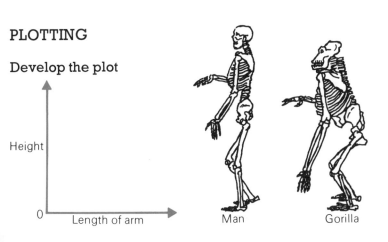

Height

0   Length of arm

Man   Gorilla

The man is taller than the gorilla, so the point on the graph representing the man will be higher than that for the gorilla.

The man's arm is shorter than the gorilla's arm, so the point representing the man will lie to the left of the point for the gorilla.

=== Exercise 4 ===

1A Copy and complete the graph above. Plot the two points and label them 'man' and 'gorilla'.

2A Copy and complete these graphs. Plot and label the points. One point has been entered for you in the first example.

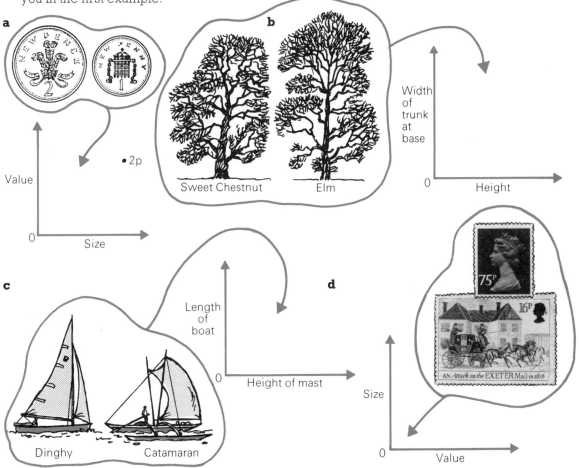

a

Value

0   Size

• 2p

b

Sweet Chestnut   Elm

Width of trunk at base

0   Height

c

Length of boat

0   Height of mast

Dinghy   Catamaran

d

Size

0   Value

GRAPHS AND RELATIONS

e

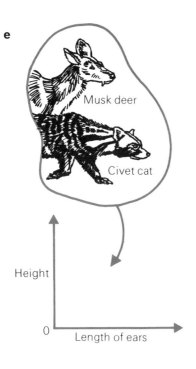

Musk deer

Civet cat

Height

0   Length of ears

f

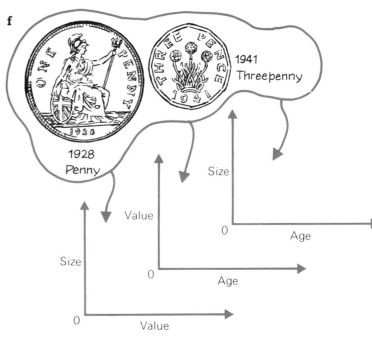

1928
Penny

1941
Threepenny

Size

0   Age

Value

0   Age

Size

0   Value

3A

2-spot ladybird   16-spot ladybird

Number
of
spots

0   Length of ladybird

Length
of
ladybird

0   Length of antenna

Copy and complete these graphs. Label the
points '2' and '16'.

These ladybirds are the same length.

4B a The 10-spot ladybird has the same
length of antenna as the 2-spot ladybird. Add the 10-spot ladybird to each
graph in question **3A**.

  b The 7-spot ladybird has the same
length of antenna as the 16-spot ladybird.
Add the 7-spot ladybird to the graphs.

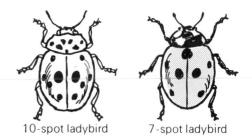

10-spot ladybird   7-spot ladybird

These ladybirds are the same length.
They are both longer than the 2-spot
and 16-spot ladybirds.

5C

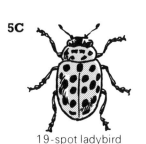

19-spot ladybird

5-spot ladybird

13-spot ladybird

22-spot ladybird

Draw a graph comparing the number of spots on these ladybirds with their lengths. Label the points '5', '13', '19' and '22'.

Sailing away

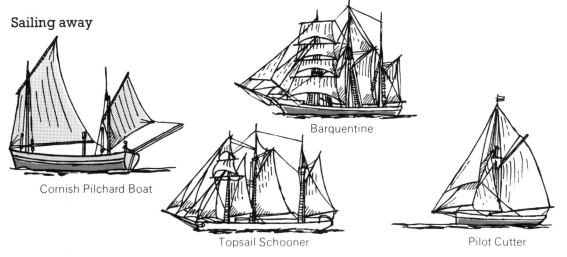

Cornish Pilchard Boat

Barquentine

Topsail Schooner

Pilot Cutter

Exercise 5

1A Copy the graph and label the points.

2A Which boat is furthest from its home port? Which boat is nearest its home port?

3B Here is a list of the boats in order of increasing height of mast:
Cornish Pilchard Boat, Pilot Cutter, Topsail Schooner, Barquentine.

Copy these axes. Plot and label four points to represent the four ships.

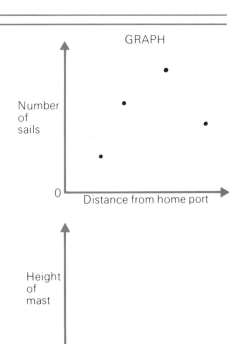

GRAPH

Number of sails

0   Distance from home port

Height of mast

0   Number of sails

Flying away

GRAPH

Length
of
plane

• Canberra

• F-15

• MIG 15

0   Height of plane

═══════════ *Exercise 6* ═══════════

1B Which plane is directly below the cloud? Which plane is furthest behind?

2C The MIG 15 is flying fastest and the Canberra slowest. Plot and label three points on these graphs to represent the planes:

Height

0   Speed

Distance
from
mountains

0   Height

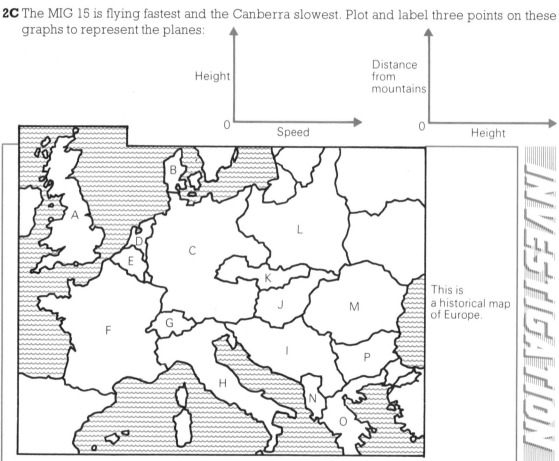

This is
a historical map
of Europe.

Two countries are neighbours if they have a common border. How many different neighbours do countries D and K have? Which of these two countries is the larger? Show this information on a graph.

Investigate the size and number of neighbours for the group of countries J, M, B and F. Draw a graph of this information.

Repeat this exercise for countries J, M, K, D and C.

Can you come to any conclusion about the size of these countries and the number of neighbours they each have?

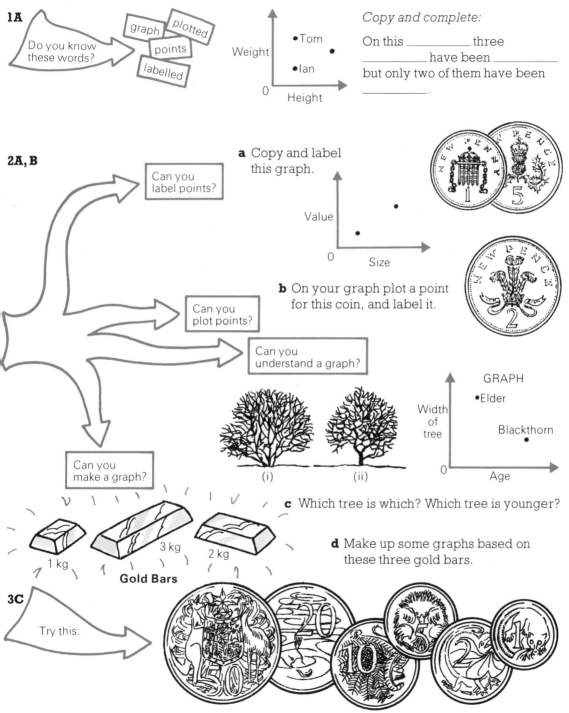

1A

Do you know these words?

graph plotted points labelled

•Tom
Weight   •
            •Ian
0   Height

Copy and complete:

On this _____ three _____ have been _____ but only two of them have been _____.

2A, B

Can you label points?

Can you plot points?

Can you understand a graph?

Can you make a graph?

a Copy and label this graph.

Value
0   Size

b On your graph plot a point for this coin, and label it.

(i)   (ii)

GRAPH
•Elder
Width of tree
            Blackthorn•
0   Age

c Which tree is which? Which tree is younger?

3 kg   2 kg
1 kg
Gold Bars

d Make up some graphs based on these three gold bars.

3C

Try this:

These coins are used in Australia. Make a size—value graph, and write a sentence to describe it. Can you think of other possible graphs?